THE CAR-DEPENDENT SOCIETY

T0264645

The Car-dependent Society
A European Perspective

HANS JEEKEL

Eindhoven University of Technology and Rijkswaterstaat (Dutch National Road and Water Agency), The Netherlands

Routledge
Taylor & Francis Group

LONDON AND NEW YORK

First published 2013 by Ashgate Publishing

2 Park Square, Milton Park, Abingdon, Oxon OX14 4RN
711 Third Avenue, New York, NY 10017, USA

Routledge is an imprint of the Taylor & Francis Group, an informa business

First issued in paperback 2016

British Library Cataloguing in Publication Data
Jeekel, Hans.
    The car-dependent society : a European perspective. --
    (Transport and society)
    1. Automobiles—Social aspects. 2. Automobile driving --
    Social aspects.
    I. Title II. Series
    388.3'42–dc23

The Library of Congress has cataloged the printed edition as follows:
Jeekel, Hans.
    The car-dependent society : a European perspective / by Hans Jeekel.
      p. cm. -- (Transport and society)
    Includes bibliographical references and index.
    ISBN 978-1-4094-3827-4
    1. Automobiles--Social aspects--Europe, Western. 2. Automobile driving--Social
  aspects--Europe, Western. I. Title. II. Series: Transport and society.
      HE5662.J44 2013
      303.48'32094--dc23

                2012044664

ISBN 978-1-4094-3827-4 (hbk)
ISBN 978-1-138-24852-6 (pbk)

Contents

List of Figures

List of Tables

Chapter 1

Frequent Car Use and Car Dependence in the Risk Society: An Introduction

1.1 Why this book?

The reason for this book is astonishment. Many activities in modern societies are not or are hardly possible without using a car. To some extent these activities have only become possible through the car. When somebody wants to transport three big tins of paint, wants to bring an old chair to her grandmother, or has to play a tennis match on Sunday morning 15 miles away from his home, a car is necessary. Only a few very creative persons can combine four activities in one day at different locations without using a car. And businessmen who need to go from one highway location to another within an hour have to use the car.

People in our time use cars frequently. One can ask whether this is always a free choice. For a whole category of activities modern Western man seems to be dependent on the car. This form of dependence is, in contrast to others, which are often defined as addiction, seen as normal. High car use is seldom seen as a problem, it is more often framed as a right! It is this situation that amazes me.

What is the reason behind this dependence? Could it be that we should talk about addiction, but an addiction that is, through its complete normality, not seen as a problem? An exciting turn-around can be seen by Miller, in the introduction to his book *Car Cultures*; 'An alien had observed that the earth is inhabited by strange creatures called cars, mainly with four wheels. These creatures are served by a host of slaves who walk on two legs and spend their whole life serving them. Cars never seem to go anywhere without at least one slave. The slaves build and maintain long and complex networks of clear space, so that cars have little trouble travelling from place to place' (Miller, 2001, 1).

In this book I will paint a broad and comprehensive picture on frequent car use and on car dependence, its driving forces, its problems and its perspectives. The book originated from my studies of the Dutch situation, but takes a broader perspective and the focus will be on Western Europe. I will present studies, ideas, facts and figures. And I will explore my themes from many angles. Frankly, I will cross all sorts of scientific domain boundaries. There is however a certain bias; this book is social science oriented, my main focus being on the social and cultural aspects of car mobility.

1.2   Central concepts

This book is built on four central concepts. The first is *car dependence*. What is car dependence? Some authors (like Newman and Kennworthy 1989, 1999; and Litman, 2002) define car dependence as synonymous with high degrees of car use. However, although the degree of car use is an important indicator, it is an incomplete indicator. Dependence points to a situation in which users do not see alternatives. Dependence has a relationship to lacking alternatives, or at least to the perception that alternatives are lacking. Car use is an indicator of the need for mobility, but not of dependence. Car dependence starts, as do most addictions, with use, but is only manifest when non-use has a negative influence on wellbeing.

Regarding car dependence one can think of a spectrum:

- from using the car
- to getting used to frequent car use
- to strong reliance on the car
- to being unable to do without a car, even when costs are high.

In this book car dependence will be described as the situation:in which a journey[1] can:

- *be impossible*
- *or only with difficulty*
- *be made by another transport mode*

Related to this definition three aspects are important. The first is the introduction of the term '*impossible or only with difficulty*'. This is a comparative concept, describing what is seen as normal in everyday life. To give an example; a bricklayer, living in an area with scarce public transport will probably use a delivery van for his work. One can argue that he does not need a van, because a moped with a small cart behind it for his tools could also be used. In the eyes of modern society, and in the eyes of almost all bricklayers, this alternative will probably not be accepted as a reasonable alternative. Furthermore, modern society sees his use of a delivery van as so normal that specific tax deductions have been created.[2]

But a bricklayer using a moped with a cart can argue that he is not car dependent. The growth of car dependence seems to be related to losing sight of convincing alternatives for car use in society.

A second aspect is the difference between the *factual and the perceived car dependence*. Factual car dependence can be measured without opinions. It is mostly not possible to travel by public transport at night. At that time, for longer distances than are possible walking or biking a car is needed.

---

1   I will throughout this book use the word journey for longer and shorter car trips.
2   As is the case in the Netherlands.

Perceived car dependence is another story. People feel dependent, but are not dependent. Alternatives are available, but are not noticed or immediately rejected. This form of car dependence is however very real in its consequences.

The last aspect is the *spectrum of car dependence*. Car dependence is a characteristic of persons, activities, times and trips and exists also at a societal level. One can safely state that the United States are more car dependent than Denmark.[3]

The next two central concepts are related; the *driving forces* behind frequent car use, and the *big societal stories*. In this book we will search for the driving forces on frequent car use. Driving forces are developments in a society. Only some of all the societal developments in modern risk societies are relevant in explaining frequent car use. For example; an important driving force in frequent car use is development which disperses activities once located within close proximity over a greater geographical area.

Driving forces form, together with expectations, emotions and abstract rules the big societal stories. These stories fixate for a certain time period the dominant views, ideas, acting patterns and codes and legitimate these. Big societal stories change through the ages, but they look relatively stable. People tend to see the big societal stories as ever remaining and continuing stories. This, however, is never the case.[4]

The final concept is the *risk society*. Frequent car use and car dependence are most obvious in the richer countries, identified here by membership of the OECD. For these countries Beck coined the concept of the risk society (Beck, 1992). Our societies, with perfectionist systems for work, housing and leisure are, in his eyes, at the same time risk societies. The risks in these societies are no longer primarily external risks (such as disasters and diseases); man-made risks are starting to dominate. These risks are 'embedded' in the expanding patterns of modern life. Focussed on car use and car dependence two questions are then of prime importance. The first is which risks can threaten frequent car use and car dependence. The second is which risks and questions do arise from this frequent car use, and this car dependence? With these four concepts we will start this study.

1.3   The structure of the book

In this book five questions will be central. These questions are:

1.   What is the current situation on car use in Western Europe?

---

3   A note about data. Car dependence is a relative concept. There are only few data to be found on car dependence. For more data can be found on car use, as we will see in the Addendum. When car dependence is at stake the use of data is always more difficult.

4   For example; the role of the 'stay-at-home' housewife, completely typical some 50 years ago, is now redundant in Western European societies.

2. Which driving forces, and which big societal stories can be seen as responsible for frequent car use, and for car-dependent behaviour?
3. What is the current situation of car dependence, and which persons and activities are more car dependent?
4. Which problems and perspectives related to frequent car use and car dependence can be seen in Western Europe in the near future?
5. Is, in the light of questions 3 and 4, a form of governance necessary and which form could such a governance take?

The book has two parts. In Part 1, Setting the Scene, frequent car use and later car dependence is analysed. In Part 2, Problems and Perspectives, the focus is on problems related to frequent car use and car dependence, and on trends inn governance in society in relation to car mobility. The book ends with a plea to take the social and cultural aspects and trends related to car mobility very serious. These trends seem at least as important as the technical or economic aspects that always dominate the academic environment on car mobility.

A central conclusion of my book is that in modern western risk societies more attention to car dependence, its driving forces, its advantages, its problems and challenges for the future is necessary.

Finally, a few words about *methodology*: In this book I will not work from a broad conceptual framework. I want to bring together all sorts of relevant literature on my two themes, frequent car use and car dependence, both in the societal context. Transparency is important; I will explain how I arrived at my conclusions. And I introduce some relevant theories such as the motility approach of Kaufmann, the mobilities approach of Urry, the time geography of Hagerstrand, the structural stories focus of Freudental-Pedersen, the transition management theories, the commons approach of Ostrom, and on the theory in governance studies on 'wicked problems', self-organisation and complexity.

PART 1
Setting the Scene

In Chapter 2 I will introduce car mobility in our modern Western European societies. My axiom is that we now live in a risk society that entails expectations and worries at the same time. Characteristics of modern western risk societies are presented, and the relationship between those characteristics and frequent car use will be explored, and some figures and facts from a number of European countries and from the EU at large will be presented.

Chapter 3 is about individual attitudes and motives behind car use. Five core motives for frequent car use will be presented, with a review of the existing literature per motive.

Car dependence is not synonymous with frequent car use. Car dependence comes into view when there is no possibility of making a trip by another mode of transport, or when this is really difficult. There is mostly no public transport at night; returning from a visit 70 miles from home is impossible without a car or motorcycle. In other situations there are alternatives, but these are not seen, or not perceived as alternatives. This can be called perceived car dependence.

In Chapter 4 I move from frequent car use to car dependence. The concept is introduced, existing literature reviewed, driving forces and structural societal stories for frequent car use identified. Also, a first quantitative current situation on car dependence in the Netherlands and the UK will be presented.

Chapter 2
Modern Risk Society and Car Mobility

Modern Western European societies and frequent car mobility seem interrelated. In this chapter I look at the core processes in modern Western societies, and I try to relate these core processes to aspects of car mobility. I will start with an introduction of basic elements of what is called the modern western risk society (Sections 2.1 to 2.5). In our societies, chances and anxieties, expectations and risks are intermingled. I will present the characteristics of modern risk societies in a scheme (2.6), and I will give some facts and figures.[1] After this introduction I will relate these basic elements, via their characteristics, to aspects of car mobility (2.7 to 2.10). This chapter should give a broad overview of societal aspects related to car use.

2.1 The risk society

Beck defines our western societies as risk societies. In his discourse three elements are central: risk, individualisation and modernity (Beck, 1992). The production of welfare in modern society is systematically combined with the production of societal risks. The logic of the distribution of wealth, that always defined society, is changed in the logic of the distribution of risks.

Richer households and richer societies try to exclude risks. They even tend to pass the risks to poorer households and poorer societies. These poorer households and societies do not accept this risk burden without argument and try to change the process of uneven burden sharing through politics, negotiations, terrorism and even war, thereby augmenting the number of risks.

Beck sees two phases in the development of the modern risk society; The First Modernity and the Second Modernity.

The *First Modernity* is related to the development of the modern nation state. Modernisation and progress were big societal projects. New risks arose as a product of these societal projects. These risks had to be mitigated and controlled in the First Modernity. That happened in different ways. A welfare state was created, with a focus on guaranteed employment, exploitation and controlling nature. Some risks were accepted as a necessity in a progressing society.

Very important in this First Modernity is the concept of functional differentiation. Integral concepts and integral thinking are then not considered very important. The common wisdom in this First Modernity is that problems, but also living areas, can be split up in different elements, and each element can in

---

1   More statistical information can be found in the Addendum, at the end of this book.

itself be brought to maturity. Mobility in this First Modernity is focussed mobility. The aim is to travel to a defined destination and to return. The great highway systems of Western Europe were built in this First Modernity.

External risks are seen as more important than man made risks. Natural disasters, floods, earthquakes and plagues in agriculture have to be fought. There is consensus on the definition of the greater risks, and hence over the necessity to control and fight these risks. It is the period where knowledge, and especially knowledge from the hard sciences, is seen as equivocal and is seen as growing steadily and fast.

Sennett (2006) calls this First Modernity the phase of Social capitalism. Stability was a characteristic of Social capitalism. Institutions were created such as unions, retirement finance and stable social dialogues between labour and capital. In Sennett's words what existed was 'the gift of organised time; you could plan your future life within the organisation' (Sennett, 2006). Stability also created predictable mobility patterns.

However, the growth in knowledge and the external effects of the arrangements chosen in the First Modernity created new risks, the risks of the modernisation processes. Control systems were built, but in Beck's words: 'these systems are part of the problem, not of the solutions, if there even is one'.

Beck's second phase, the *Second Modernity*, comes into view at the moment when developments cross the boundaries of the nation states. Risks become more man-made, cannot be controlled any longer, but are inherent to the existing lifestyles and arrangements. People are well aware of these risks. A society that accepts manmade risks, and makes these risks a theme in discussions and policies can be seen as 'reflexive'. Beck calls this the phase of Reflexive Modernisation (Beckmann, 2002; Beck, Bonss and Lau, 2003)

In this Second Modernity we see the globalisation of capital and economic life, we next see phases in individualisation, flexibility in working arrangements, work without security, without life-long perspectives. And we see a worldwide ecological crisis. Mobility loses its predictability; criss-cross patterns grow, trip-chaining becomes normal. The car becomes an integrator between different 'areas' important in modern life; the work area, the shopping area, the leisure establishments and the home.

The Second Modernity is, in Beck's vision, synonymous with permanent transformation, unpredictability, chaos and permanent change in just accepted realities. Man-made risks become central, and there is a whole portfolio of such risks. Originally Beck focused on technical and environmental risks, related to his experiences with environmental disasters like Love Canal and Chernobyl. Nuclear energy, toxic waste and environmental degradation then come into view. At a later stage he also focuses on the risks of the biosciences such as cloning and artificial life and his attention spreads towards wars, terrorism and global insecurity. Recently the risks of global finance have become very clear.[2] All these

---

2   In regard to the global financial crisis, see Beck (2008) and Geldof (2008).

risks are addenda of essential activities in modern societies like energy production, economic development and greater influence over your own living conditions.

All these risks have in common the fact that there is no easy answer for the question of which effects will be really dangerous. Dangerous is a concept open to debate in Second Modernity. Is nuclear energy a huge risk, or cloning? We try to manage risks, and risk science or risk management are popular. But even if we reach consensus on the definition of the risks, these new risks are too big and too global to be controlled with the instruments created in the First Modernity.

Society at large feels this, but continues to ask for better protection. People know that these risks are integral to the arrangements of their lives, but they are quite happy to deny this wisdom when they spot a leader. For leaders this ambivalence is difficult. They cannot use their old rituals. They have to produce safety and security, although they know that safety and security cannot be produced on a large scale. Their language seems unfit for any societal accepted purpose. The language of the First Modernity does not function anymore in the Second Modernity.

2.2   Differentiation and a greater scale

An important characteristic in modern risk society is the unbundling of activities. Until relatively recently the locations for various functions were situated in proximity, or even interwoven. Local industries were concentrated in the same area as housing and local shops. Modernisation has created separate spaces, separate locations for several functions. We now have industrial areas, shopping malls, and housing areas without much other activity.[3] Proximity is no longer necessary; the car plays a role connecting all these locations.

In modern risk societies we also see forms of 'social unbundling'. The spectrum, the variety of living patterns, has grown. Individual choice on how to arrange your life has become more important. Working and framing from clear societal models becomes difficult. To quote Beck,

> but what is a 'household' nowadays, economically, socially, geographically, under conditions of living apart together, normal divorce, remarriage and transnational life forms? High mobility means more and more people living in a kind of place-polygamy. They are married to many places in different worlds and cultures. Transnational place-polygamy, belonging to different worlds, this is the gateway to globality in one's own life. (Beck, 2002, 24)

Modernisation becomes more problematic, according to Beck:

> The continued technical, economic, political and cultural development of global capitalism has gradually revolutionized its own social foundations. In the

---

3   Analysis for example in Harvey (1989).

> transformation from the first modernity, that was largely synonymous with the
> nation-state, to the second modernity, the shape of which is still being negotiated,
> modernization ends up stripping away the nation- and welfare state, which at
> one time supported it, but later restrained it. (Beck, Bonss and Lau, 2003, 2)

This can be an intermediate phase. Beck expects that we are growing towards a
cosmopolitan society. He sees the loss of attachment towards a place, a specific
location, the loss of proximity; 'It means; loosening and transforming the ties of
culture to place … . Sociability is no longer dependent on geographical proximity.
It thus becomes possible for people who live isolated from their neighbours in one
place simultaneously to be tied into dense networks stretching across continents.'
(Beck, 2002, 31; and Beck and Sznaider, 2006). In *The Cosmopolitan Society and
its Enemies* (2002) he pictures the broader vision. He sees a growth towards such
a society, but; 'cosmopolitanism lacks orientation, perhaps because it is so much
bigger and includes so many different kinds of people with conflicting customs,
assorted hopes and shames, so many sheer technological and scientific possibilities
and risks, posing issues people never have faced before' (Beck, 2002, 20).

2.3   The urge for flexibility

In the economic sphere we already face a global world. Financial markets are arranged
in a combination of computer-driven manoeuvres, based on mass psychology, with
unexpected turbulence from the great complexity of the many financial interventions
at the same time. The world, in global concurrence, rewards cheaper regions, or
regions with high value added. 'Lean Production' is a key message. Low production
costs demand subcontracting, outsourcing, rationalisation and downsizing.

Manuel Castells uses for this complex the term 'space of flows' (Castells,
2000); streams become globally oriented, flexible, changing daily. Older standards,
experiences, values and institutions have been replaced by fluid arrangements.
Mobility, or rather, the capacity to keep moving, is now a key factor in societal
success. Two elements become essential: flexibility and versatility.

However; the problem is that for most people their experiences and their
creation of significance are still locally-based, at least certainly not globally-
based. Scaling up comes at a cost. The core processes in the economic sphere have
no relation with the platforms and areas where social significance can be created,
and where political power can be used. Castells (2000) signals that there is no
continuity between the power constructed in the global economic networks, on the
one hand, and the logic of political representation on the other. Nation-states seem
unable to create countervailing power toward the financial-economic processes
that are in essence without any morality. It is now impossible for most nation states
to define their own monetary policy, and to decide about budgets, or to organise
trade and collective taxes without long and difficult negotiations with other nation-
states. National governments face a loss of power, but try to create new power at a

higher level. At the same time they ask their voters to follow them, to be flexible, and to respect the laws of global finance without morality. In doing so, national governments are facing the loss of contact with their voters, who still expect some safeguarding against too much fluidity and flexibility.

We are entering a phase of premature globalisation, with a split between global-organised economic power, and local- and pastoral-organised political power. To cite the sociologist Bauman, 'the global space has become a frontier land of sorts: a kind of 'extraterritorial territory' without binding laws and rules of conduct, a battleground of undefined or shifting/drifting frontlines and floating coalitions' (Bauman, 2001). Politicians are in his eyes not able to create 'civilisation' in the financial-economic processes. In Bauman 's world, progress has a specific connotation, 'progress is thought about no longer in the context of an urge to rush ahead, but in connection with desperate efforts to stay in the race' (Bauman, 2001).

In *The Corrosion of Character* (1998), Sennett describes how people in their working life have to be more flexible than ever. That seems to be a necessity in the New Capitalism, globally organised and aiming at short-term time profits, related to shareholders' value. New Capitalism asks for adjustment to new situations all the time, to intense flexibility. Stable institutions and practices are barriers in the new economic order.

People see companies come and go, see mergers combined with efficiency cuts, meaning loss of employment possibilities all the time and feel that 'the economy' is the hidden force shaping their lives. They also notice the other side of New Capitalism; countries and neighbourhoods in western cities not included in the global money-making networks where nationalism, xenophobia and terrorism find their basis. All this creates widespread anxiety. In Castells' worldview, people living is these countries and neighbourhoods want to have access to welfare, but the politicians and citizens of the richer countries and neighbourhoods are fierce 'to keep barbarians of the impoverished areas of the world outside the gate' (Castells, 2000). Poorer countries, poorer neighbourhoods, and especially public spaces in these areas are becoming less safe.

Individualisation and the need to move with the economy create in Bauman's view a world in which people feel urged to greater flexibility with less social security, which leads to a loss of trust in other people. He describes in *Liquid Fear* (Bauman, 2006) incompetence as a major source for modern anxiety. People feel incompetent because they miss the power to adjust all the time to new flexibilities, and they see that leaders and governments are not supporting them. They sometimes see the future as 'a walk in thick fog'.

New Capitalism asks people to develop a 'flexible self' and to watch carefully for problems and opportunities. Stability is not created in this New Capitalism. To introduce Sennett, 'the flexibility that new capitalists love offers no guidelines for living a good and normal life, and is also not able to give those guidelines' (Sennett, 1998, 160). And he concludes, little prophetically in the light of the credit crisis, 'an arrangement that gives no reason for people to care and to care about each other cannot retain its legitimacy very long' (Sennett, 1998, 160).

In the Anglo-Saxon world particularly, the birthplace of New Capitalism, there has been a loss of 'rituals, in which social integration can be lived'.[4]

2.4   Anxiety and uncertainty in the risk society

A risk society produces at the same time anxiety and insecurity, and expectations and chances. The equilibrium between these four elements seems crucial.

The greater individual freedom, the richer variety, and the loss of standard behaviours cause, at a societal level, the disappearance of a sense of direction. At the personal level feelings of anxiety and insecurity arise. To quote Boutellier, criminologist and author of *The Safety Utopia*, 'In a risk culture moral discomfort generates a need for safety' (Boutellier, 2002). A more fluid lifestyle is created, with a loss of long standing orientation marks. Boutellier again; 'our culture is not a culture of learning and knowledge accumulation, but of discontinuity, of forgetting, and starting all over again' (Boutellier, 2006, 38).

An amalgam of rising expectations, discomfort, a wish for transparency and a desire to mitigate and control risks of modern life creates a strong plea for safety, for getting rid of anxiety. A study from the SCP (the Netherlands Institute for Social Research, 2006b) noted that 15 per cent of the Dutch adult population feels more or less unsafe. Feelings of insecurity are highest at night, outside the known living spaces, in areas where many non-Dutch citizens live, and in neighbourhoods that appear to be corrupt.

Where can we situate the reason for these perceptions of insecurity, which are certainly not matched by the criminality statistics in the Netherlands? A first explanation has to do with the focus in the media (see Altheide, 2002). In most western countries journalists report with great frequency about the most risky and violent incidents, and they report about these incidents several times. People love to read their stories, and thus create for themselves feelings of insecurity.

Two British scientists, Durodie and Furedi, offer a more elaborate explanation. In *The Limitations of Risk Management* (2005), Durodie, the former director of the British International Centre for Security Analysis, criticises the start of discussions about risks, insecurity and safety. The focus is on managing and mitigating risks and far less on the use of our human capacities to organise our lives in a more controlled way. In his words 'to take a risk' has become 'to be at risk' (Durodie, 2005, 14). Modern societies are very defensive about risks, 'we do not have a risk society but a risk perception society' (Durodie, 2006, 2).

Durodie's approach focuses on the term 'resilience'. Resilience has been lost. We have become anxious, and the feeling of purpose and will seems to be on the loss in western societies. Why? In Durodie's opinion, 'key element in shaping our

---

4   A term of Van Dinten, former Head of Strategy of the Dutch Rabobank. Van Dinten continues, 'we created an infrastructure in which the possibility to feel the need to cooperate has been blown away'(Van Dinten, 2006, 548).

perceptions of risk and the management of most policy issues today is a sense of isolation and insecurity that affects every layer of society'(Durodie, 2005, 16). People living in each other proximity do not know each other, are socially not interrelated. They do not keep an eye on each other. They go their own ways. Durodie signals that resilience will grow, and hence feelings of insecurity and anxiety will diminish, when we know better – in connection with our fellow human beings – what to strive for, who we want to be, and what we are aiming at. The 'catastrophic absence'[5] of investments in joint objectives and goals makes modern people far more vulnerable than necessary. Locke heads in the same direction. The central thesis in *The De-voicing of Society* (1998) is that the price paid for greater freedom of movement for modern man has been a growing anonymity in the social spheres. Small isolated private introspection leads to framing everything unknown and social as a potential or actual risk.

Durodie finally argues that the best approach to risk management is to restore the connections with our fellow human beings. Competent risk management needs trust, and we have lost too much our trust in our fellow human beings.

A sort of 'infrastructure of threats' (the term used by Beland, 2005) has developed. Many people and many activities or circumstances seem to threaten us (Cebulla, 2008). This spectrum of threat is a central theme in *Paranoid Parenting* (2001) by the British sociologist Furedi. Parents nowadays are very concerned about the safety of their children. In the eyes of British parents almost everything around their children can create risks. The anxiety about the fate of their children is higher outside the parents' zone of influence. What happens at school, in the public space, are the games they are supposed to play not too risky? (Furedi, 2002). Lots of parents keep their children in controlled spaces, mostly in their homes. Many parents do not dare to let their children go somewhere by themselves,[6] they feel the urge to accompany them everywhere, even when they are older.

Furedi (2004) also sees the reason for this parental behaviour in the loss of parental solidarity. The trust that other parents will watch and take care when something happens to your child seems lost. Parents feel themselves alone, and this leads to insecurity. Pedagogues know, however, that children play best when their parents are not around. Delfos (2004) signals in *Werken is nepspelen* that playing in public spaces, and on the streets is becoming more difficult and less accepted while 'a child learns justice best on the streets and on playgrounds, in open and active play with other children'(Delfos, 2004, 133).

---

5   Durodie states; 'presumably, people are prepared to risk their lives fighting fires or fighting a war, not so that their children can, in turn, grow up to fight fires and fight wars, but because they believe that there is something more important to life worth fighting for. It is the catastrophic absence of any discussion as to what that something more important is, that leaves us fundamentally unarmed in the face of adversity today' (Durodie, 2005, 18)

6   Furedi mentions the head of a British primary school; 'some girls do not even have a raincoat, they are transported everywhere, all the time, by car' (Furedi, 2001, 10).

Modern parents do not have fewer acquaintances than earlier in history, but they do have fewer acquaintances in the vicinity of their home. Where education takes place, and that is essentially local, and in normal life, people feel alone. To cite Beck; 'each of us is expected to seek biographical solutions to socially produced troubles'.

2.5   Expectations and chances in the risk society

Until now my analysis was a little pessimistic. There is, however, another side to the modern risk society. People are thriving, feel that they have more opportunities than generations before, and have many expectations.

Anthony Giddens sees three main changes in modern society: globalisation, information technology with the inherent necessity to define positions towards the ambivalent possibilities of modern science, and, lastly, the transformation of daily life, with less importance for the nuclear family in relation to the growth of other types of households (Giddens, 1991). Giddens does not see modern people as helpless individuals. In his structure theory he asks whether people or social forces are behind the steering wheels of change. Modern people create our society, but also feel restricted by modern society. We now have access to all sorts of information that give us the possibility to reflect on our own interventions and activities. In the meantime we see ourselves confronted with dangers, risks and unforeseen effects of our interventions.

Giddens notices that modern people have become detached from the restrictions of space and time. We can experience other times, and other places than the ones we are in now and we have to accept that professionals structure parts of our life. In Giddens' words 'trust in a multiplicity of abstract systems is a necessary part of everyday life today, whether or not this is consciously acknowledged by the individuals concerned' (Giddens, 1991). Modern people will give that trust, but are not convinced of the positive use of the techniques that have been created. Many possibilities, and at the same time many dangers, create insecurity in our societies. Giddens' plea is to introduce 'life politics', the politics of self-actualisation. Modern people have to work themselves towards happiness, have to define what they can do, what they wish, and what they want.

In emphasising the possibilities for man to influence his circumstances Giddens differs from many society watchers. People do not have to waste time in empty spaces, but can actively take their life in their own hands. A certain level of autonomy is essential, and hence a certain level of education.

The Dutch social scientist De Beer looks at a cornerstone in the work of Beck and Bauman. Using data from the Dutch SCP he examines whether our modern society is indeed more individualised. He presents his conclusions in *How individualised are the Dutch?* (2007). After mentioning that Beck's and Bauman's descriptions are too vague for data research he dissects individualisation into three sub-processes:

1. Loss of traditions; operationalised by the membership of traditional organisations.
2. Emancipation, operationalised by the question whether attitudes and behaviour have become less predictable.
3. Heterogenisation, operationalised by growth in variety of choices.

He concludes that the Netherlands have become less traditional. Heterogenisation did occur, but there are times with greater, and times with lesser, variety of choices. The most remarkable conclusion is that attitudes and behaviour are still predictable, looking at categories like age, gender, and social class.

The Flemish sociologist Elchardus continues in this direction. In *We lopen een culturele revolutie achter* he concludes that individualisation is not growing very fast (Elchardus, 2004). Societal splits are still there, but old splits have been replaced by new ones. Flexibility in working hours, in shopping hours, is growing slowly. A 24-hour economy did not arise. Elchardus sees the role of the media as a new element. The media structure our life. We can, with Giddens, choose for ourselves, but pretty often the same choice is made in each segment of society. These segments, functions of income, education, and background are still defining these choices. Socialisation processes create those segments, the new 'social classes'. Elchardus writes,

> *de laatste 50 jaar werden een aantal instellingen ontwikkeld zoals het algemeen onderwijs, de massamedia, vooral de televisie, de reclame en de symbolische organisatie van de distributie van de consumptiegoederen, die een langdurige, diepgaande en voortdurende vorming en beïnvloeding mogelijk maken. De leden van ons type samenleving worden nu, van kindsbeen af, beïnvloed in hun kennis, vaardigheden, opvattingen, houdingen en smaken.* (Elchardus, 2004, 13)

> in the last 50 years new institutions were created like the general education, the mass media, especially the television, the advertisement and especially the symbolic organisation of the distribution of goods. All these institutions influenced and created learning loops. The members of our societies are now from childhood on, influenced in their knowledge, skills, attitudes and tastes. (author's translation)

Elchardus does not present a picture of lonely individuals, following their own solitary road but many people taking the same road.

In our modern risk society people have many expectations. The Dutch sociologist Van den Brink focused on these expectations in *Hoger, harder, sneller... en de prijs die men daarvoor betaalt* (2004). He presents a time perspective with many data and identifies three important groups in modern society. Nearly 30 per cent can be called the 'threatened citizens'. They can be characterised by low income, low education and distance from modern societal dynamics. They are not able to follow the urge for flexibility. At the other end of his spectrum are the 20 per cent 'active citizens'. They mostly have higher incomes, higher education and show great interest in civil society. They are critical about most politicians. The biggest group, 50 per cent, is the 'accepting middle class'. They accept the modern world, but do not initiate new practices themselves.

Central in Van den Brink's vision is that growing prosperity gives rise to growing expectations. Dissatisfaction arises when modern living conditions cannot meet the standards related to these growing expectations. It should be clear that we talk here about another type of dissatisfaction than Beck, Sennett or Bauman. He clarifies his vision for a few areas of life. Partners expect more from their relationships than ever, the burden of affective problems is growing. In the work sphere, demands from employers towards employees are growing, and employees have higher expectations of their careers. It is rather difficult to meet these higher expectations, so a price is paid for all these higher demands (Van den Brink, 2004, 20); more divorces, more stress. Van den Brink sees moral problems and confusion only among the threatened citizens.

However, stress is in his vision a new and important category. Women in particular have to combine activities in different domains, in rather tight timeframes 'the housewife – spending a great deal of her time in and around the house, and being able to follow her own time schedule – knew margins. These margins disappeared, and the former housewives' tasks are being spread over more and more busy people, thus creating stress in society' (van den Brink, 2004, 26).

Van den Brink defines a process of growing expectations. Norms and standards for social interaction rise. We become more sensitive to inconvenience, to nuisance, and to risks and we have higher, mostly implicit, demands on interaction. Dissatisfaction is growing, and there is no single body able to coerce these higher standards. Modern society creates higher norms for living together, but most citizens also accept that these norms will probably not be met. Many threatened citizens however cannot cope with this situation. They want more clarity and security than modern society can offer, and populist parties are voicing their problems.

2.6   Facts and figures on modernity and car mobility

We have tentatively discussed elements in modern risk societies. I will present a summary of the characteristics in Table 2.1. The important question is how these characteristics relate to car mobility.

The car has a double bond with the development of modern risk societies. The car helped to create modern society. Frequent car use is, then, a consequence of the characteristics of modern society. Car dependence is the next step after frequent car use.

Car ownership has grown in recent decades, and is still growing. In most Western European countries, there are now more households with two or more cars than households without a car. From Table 2.2 it can be concluded that 18 to 25 per cent of all households have no car. Forty-three to 55 per cent of all households have one car and 20 to 32 per cent of all households own two cars. More than two cars can be found in 2.6 per cent to 6 per cent of all households.

The rule of thumb for the selected seven countries[7] is 21 per cent carless households, 49 per cent one car households, and 30 per cent households with two or more cars.

Table 2.1   Characteristics of the modern risk society (core authors shown in italics)

Characteristics modern risk society	Societal characteristics that have become less important
Abstract systems *Giddens*	Authority
Wish for simplicity and authenticity *Boutellier, Sennett*	Easy readability of society
Moving forward without direction *Bauman, Boutellier*	Progress and direction
Expectations and wishes *vdBrink*	Waiting
Implicit expectations in the social sphere *Beck, Bauman*	Social connections
Differentiation in lifestyles, no common ground *Elchardus, Beck*	Class thinking
Scepticism about science and technology *Beck*	Belief in science and technology
Leading roles for the media *Durodie*	Leading roles of institutions
Fears, feeling vulnerable, seeking security *Beck, Durodie, Furedi*	Resilience and composure
Flexibility as a necessity *Castells, Sennett*	Recurrence as an asset
Active leisure as an asset *Boutellier*	Relaxation
Everything on appointment *Furedi*	Just walking in
Geographical spreading out of activities *Beck*	Investments in proximity

---

7   These seven countries (in fact; six countries plus the northern part of Belgium) have relatively recent and comprehensive National Travel Surveys. For the surveys used, see the Addendum, and note 2, page 131.

Table 2.2   Ownership of cars, by households (%)

Country	No car	One car	Two cars	More than two cars
Germany	18.0	53.0	23.0	4.0
United Kingdom	25.0	43.0	26.0	6.0
France	19.0	45.0	32.0	4.0
Switzerland	18.8	50.6	25.1	5.4
Flanders	18.2	53.6	24.7	3.5
Sweden	25.0	52.0	20.0	3.0
Netherlands	20.9	54.9	21.6	2.6

Source: Different National Travel Surveys, different years, see note 2, page 131.

Car ownership is still growing quickly in Western Europe. In the Netherlands car ownership in 1995 was nearly 5.7 million cars, compared to 7.6 million cars in 2009, a growth of 33.3 per cent or around 2.4 per cent yearly. Car density in the Netherland went from 390 cars per 1,000 inhabitants in 1995 to 460 cars per 1,000 inhabitants in 2009. In the United Kingdom between 1994 and 2008 the number of cars went from 23 million to 31 million, a growth of nearly 35 per cent, or around 2.3 per cent yearly. And in France from 1994 to 2008 the number of cars grew from 26.4 million to 32.7 million, a growth of 24 per cent, and around 1.7 per cent yearly.

The number of cars per 1,000 inhabitants is called car density and data for the whole of Europe are available (Eurostat, 2009). Starting with Western Europe, many countries had, in 2008, a car density around 460–500 cars; Sweden (466), Norway (464), United Kingdom (464), Belgium (481), France (485). The German-speaking countries have somewhat higher densities; Germany (503), Austria (515), Switzerland (530) and, particularly, Luxemburg (681!). When we look at the southern part of Europe we see Spain with 489 and Italy with 609. More to the east we see mostly lower densities; Poland (422), Estonia (412). Latvia (411), Czech Republic (426), Slovakia (286), Romania (187), Bulgaria (310), Hungary (333) and Croatia (350). Only Lithuania (496) and Slovenia (520) have high car densities.

The conclusion from these patterns of car growth and car density should be that Western Europe and Southern Europe will soon reach the situation where a car is available for every second person in the population.

In *Panorama of Transport* (2009) the Statistical service of the EU, Eurostat, presented the average daily distance per mode of travel for all European countries. The data for this is presented in the Appendix in Table A7 and is for cars and public transport (separate bus and coach, railways, trams and metro), but not for slow modes.[8] A rather well defined picture can be seen; most Western European countries are high on car use and high on the number of kilometres travelled per day. Most Eastern European countries (with the exception of Slovenia and Lithuania) are not yet strong on car use and are lower on the number of kilometres travelled per day.

Although the basic data differ somewhat from the National Travel Surveys, a rather well-defined picture can be seen. Most Western European countries are in the left and higher part of the table. These countries are high on car use and high on kilometres travelled per day. Most Eastern European countries (the exceptions being Slovenia and Lithuania) are in the right and lower part of this table. These countries are not yet strong on car use and lower on kilometres travelled per day. What has happened to explain all this motoring?

Modernisation is essentially a process of structural social differentiation. What formerly belonged together is now spread out, and each element develops in its own defined niche. Differentiation can, however, only exist with a parallel process of integration. The German transport scientist Rammler says in *The Wahlverwandtschaft of Modernity and Mobility* that mobility takes care of the spatial integration of the social differentiation. Transport infrastructures, and certainly the recent car system, are 'both skeleton and nervous system of modern industrial growth society' (Rammler, 2008, 67).[9] The car, being able to come everywhere, is essential for the social integration of modern man.[10] Social differentiation continues; functions, tasks and services are spread further away, and each time the car has to connect all these elements. This means still greater car distances, and it means more reliance on the car to reach all the essential elements and services of modern life 'transportation is a force that holds the modern world together while driving it apart' (Rammler, 2008, 70).

A general picture can now be shown on two levels. First at the *general level*: Mobility in Western Europe is, in recent years, growing slowly after a period of rather strong growth between 1975 and 2000. Growth comes mostly from growth in population, and less from growth in distance travelled per person. This last form of growth is diminishing rapidly. *Within the total mobility (globally 36 kilometres per person per day) only the car kilometres are still growing.* Car kilometres account for 79 per cent of the total kilometres travelled and this share will grow, over the next decade, to over 80 per cent. Public transport now has 12.5 per cent of the distance travelled, and the slow modes account for 7.5 per cent, somewhat more cycling than walking. The car share in journeys (globally 3.15 mobility journeys per person day) is lower, around 56.5 per cent, public transport covers

---

8   This means that, in most countries, in their Travel Surveys car percentages of 78 per cent with small modes end in the 1980s in the Eurostat data without the small modes.

9   Broader introductions to the role of the car in modern society can be found in Lyons (2003b) and Beckmann (2001).

10   An interesting study on this subject is by Heine and Mantz; *Mobilitat und Grenze des Autoverzichts* (2000); car use related to the lives of German women in Hanover.

10 per cent of the journeys and 34.5 per cent are made by slow modes. Shorter trips are made walking or cycling, and for journeys over 7 km the car dominates. Although the distance travelled is not growing much in Western Europe, the number of cars still is still growing relatively quickly. While distance travelled per car is diminishing, the number of cars per household is rising. *Cars become individualised consumption articles,* like clothing. Also diminishing is the number of passengers per car. Driving means; driving alone.

At the *household level* nearly half of Western European households have one car. Thirty per cent have two or more cars, and this share is growing, and 21 per cent of the households – mostly single households, the elderly, and the very young households – do not own a car. The majority of families with children now own two or more cars. Driving license density (81 per cent, men 87 per cent, women 74.5 per cent) is a little higher than car density (79 per cent). In one tenth of the households without a car someone can drive.

Car driving costs 15 per cent of net household income. This share is more-or-less constant. Poorer households have fewer cars, and travel shorter distances in general. The same holds true for less educated households. Reliance on a car is highest in rural and peri-urban areas.

Car mobility is least important for households in Western Europe's major cities where public transport has an important share. The car dominates in work related mobility (with 70 per cent of the journeys and 75 per cent of the distance). Most work travel is driving alone. For leisure and shopping the car is used primarily for longer distances and education is the least car-dependent motive. Because cars have more passengers in leisure traffic this motive is the winner in total car distance travelled. An important newcomer among the motives is escorting, an indicator that the car is becoming a necessity in the fulfilment of the duties of daily life.

In the following paragraphs the different appearances of mobility will be shown in relation to the characteristics of modern risk society discussed. A kaleidoscopic picture of mobility in the Second Modernity will arise.

Table 2.3   Characteristics risk society and mobility

Characteristic risk society	Appearance in the mobility domain
Differentiation at a greater scale (2.2)	Peri-urbanisation Trends towards highway locations (2.7)
The urge for flexibility (2.3)	Mobility in networks, and motility (2.8)
Anxiety and uncertainty (2.4)	Mobility patterns related to children Mobility patterns in leisure (2.9)
Expectations and chances (2.5)	Idem (2.9)

2.6 Peri-urbanisation and the rise of the highway culture

In this Section I will present two examples of the differentiation, mentioned in Section 2.2 on a greater scale, that are car mobility related. I will start with the development called peri-urbanisation.

The urban field has spread in recent decades. Peri-urbanisation is urbanisation at a relatively great distance from the central city. The term was coined in France, where people working in Paris now live in villages and small cities 100 kilometres from the centre of Paris centre in '*la couronne francilienne*'. The peri-urbanised territories are now the fastest growing territories in France in terms of population and housing stock (Massot and Roy, 2004). Newcomers are mostly higher officials and freelancers. They are very dependent on car travel to reach their workplaces, as there is mostly little public transport available.[11]

Motte-Baumvol looked at activity patterns in *Les populations peri-urbaines face a l'automobile en grande couronne francilienne* (2007). Car dependency is greater in villages where there are no services but even when there are services car use is on a higher than average level than France as a whole. Peri-urbanisation is, in the eyes of Korsu and Massot (2006), a threat and a preference. People like to live there, but they feel vulnerable as well. The distance to workplaces, and sometimes also to leisure, is very great and they have to face congestion near Paris, or on a broader scale, near other cities, and they fear higher petrol and oil prices in the future.

It is not only these newcomers that are located in the peri-urban areas. The longstanding population, mostly with far lower incomes, is still living there. Fifteen per cent of peri-urban households and clearly a greater part of the long standing families do not own a car. They are seen by Rouge and Bonnin in *Les 'captifs' du peri-urbain 10 ans après* (2008) as 'captives'. Rouge and Bonnin concluded that living in these areas is not easy for these captives, because of higher prices (their small shops closed and were replaced by supermarkets at greater distances), fewer acquaintances and the growth of structures of car space. Motte-Baumvol (2007) saw three practices within carless households; help from the family (in French called '*covoiturage*'), living in a small area, and trying to use public transport. A part of the non-car households moved nearer to Paris. Living in a smaller area was clarified; 60 per cent of the journeys made by these households covered a distance of less than 1 kilometre, compared to 35 per cent of all peri-urban households. From this French literature it becomes clear that the process of differentiation on a greater scale is not a neutral process. There are winners and losers but even the winners feel vulnerable sometimes and hope that conditions on their daily commute will not change for the worse.

---

11   On the situation in Wallonia, the French speaking part of Belgium and an area with much peri-urbanisation, analysed in relation to the Danish and Swiss situations, see Halleux et al. (2002).

The other example is highway locations. In Europe in the last two decades many *highway locations* have been developed. Along the highways in a great number of countries you can now visit office parks, business parks, factory parks, industrial zones and housing sites. Not all countries have many highway locations. It looks like there are country-specific situations. For example Spain has many highway locations, while in Germany along the *Autobahnen* there are practically no highway locations. Belgium and the Netherlands have many highway locations compared to England or Denmark. France has an intermediate position, with highway locations situated more-or-less exclusively along the urban stretches of the highway network. In the Netherlands highway locations are a recent phenomenon (Hamers and Nabielek, 2006).

Office parks are found more in the *Randstad* area, while factory parks and industrial zones are concentrated on highway locations outside the *Randstad*. Dominant activities on highway locations are business-to-business services, logistics, lower value-added offices, and factories and workplaces that need space. At the moment 23 per cent of the Dutch workforce works at a primary highway location (defined as less than 1,000 metres from a highway exit) and a further 18 per cent of the workforce works at a secondary highway location (less than 2,500 metres from a highway exit). By about 2020 around half of the Dutch population will work at a primary or secondary highway location (Jeekel, 2011, 391).

Until 1990 the Dutch government was very reluctant to develop highway locations. The government even introduced a policy, the so called A, B, C policy, to control the location of offices, businesses and companies. This policy clearly failed (see Martens, 2000) and now highway locations have three distinct geographical appearances.

Factory and company parks and industrial zones have been created by municipalities. Municipalities support their local entrepreneurs. When an entrepreneur considers moving his factory necessary, and asks for support for building expansion, most Dutch municipalities are quite willing to buy agricultural ground in order to support the local economy. The alternative, restructuring already existing industrial sites that have become old-fashioned in order to create capacity, is considered too difficult by most municipalities. This municipal behaviour has two effects. First; fast urbanisation of cheap agricultural land, mostly situated along highways. No further infrastructure is created, there is mostly no public transport and the entrepreneur usually builds non-sustainable buildings, with average lifetimes of 12 years. Secondly, economies of scale are missing. Each municipality follows the same practice, resulting in a reservoir of ground for new industrial developments in the Netherlands (see Olden, 2010).

The market for *office space* in the Netherlands is essentially a commercial estate market. There is a preference for offices in inner cities but because of parking problems many offices are located along highways, mostly in suburban locations in the *Randstad* area (see for figures, Bureau Louter, 2007). The market for office space is a fast market; older offices are easily left for newer ones. Six to eight per cent of Dutch office space stands empty (Remoy, 2010).

In recent years *big housing areas* have been developed adjacent to highways. The idea was to strengthen the city to which these housing areas belonged. However, this has been a miscalculation. The fact that the new neighbourhoods are situated in the vicinity of a highway means that the new residents drive to all the locations within easy reach. The city they formally belong to is just one of many destinations. They become footloose, and they become car dependent. The new residents of these neighbourhoods situated near highways have the highest car use in the Netherlands (Snellen, Snellen, Hilbers, Hendriks, 2005, 66 and Table 16).

Sixty per cent of all work in highway locations is part of a built-up area. The other 40 per cent consists of solitary locations often in the countryside. These locations mostly have no public transport and are rather difficult to reach by bike (Jeekel, 2011). In most areas new highway locations create extra congestion (see Hilbers, Snellen and Hendriks, 2006).

After much criticism (see VROM Raad 2006) the Dutch government pushed for a more concise policy on industry and company parks, and on industrial zones. A commission was created in 2008 that suggested making planning for industry and company parks a regional activity (see Taskforce, 2009). Municipalities do not want to follow this suggestion (VNG, 2009).

From a viewpoint on sustainable development the role of the Dutch municipalities is questionable. Among the population is a dislike of working in these highway locations (DHV/Smart Agency, 2007). The fact is, however, that these locations are still being created. In fact almost 10 per cent of Dutch households now have at least one member living and working along the highway. In 2020 this will be almost 14 per cent (Jeekel, 2011, 400). Relatively late, but in the Netherlands we now speak of a highway culture, with as its newest development being the growth in leisure locations such as cinemas and discos.

2.8 Networks and motility

Theory building about physical mobility in modern societies has only recently developed. It was a more-or-less forgotten subject. A core of activities is found around the journal *Mobilities*. Most social scientists on mobility see this journal as a useful network opportunity and British, German, Swiss and Scandinavian scientists dominate. There are only a few connections to the Latin-speaking world. France has built its own social science knowledge on mobility through the work of Massot, Orfeuil, Genre-Grandpierre, Motte-Baumvol, Halleux and Dupuy,[12] to name a few. Core themes in France are peri-urbanisation, the role of time and speed in mobility, and gender issues.

---

12 Dupuy (1999) was the first to publish a book on car dependence in the Latin speaking world; *La Dependence Automobile*. See also his recent *Towards Sustainable Transport*, 2011.

Mobilities has as central themes; the moving of people, goods, capital and information on a large scale, local processes around transport and traffic, interactions in the public spaces and transport and mobility issues related to day to day life. Note that the focus is on migration as well as on urban mobility.

Networks, flexibility and car mobility

In the eyes of Urry, the central theorist of the Mobilities group, car mobility is a hybrid compilation of people, machines, roads, buildings, signs and cultures. The 'structure of auto space' has, as a first step towards Second Modernity, asked people to orchestrate their mobility and their social contacts. In *The 'System' of Automobility* he states:

> the urban environment has 'unbundled' territorialities of home, work, business and leisure that historically were closely integrated, and fragmented social practices in shared public spaces. Automobility divides workplaces from homes, producing lengthy commutes into and across the city. It splits homes and business districts, undermining local retail outlets to which one might have walked or cycled, eroding the centres, non-car pathways and public spaces. It separates homes and leisure sites only available by motorized transport. Members of families are split up since they live in distant places involving complex travel to meet up even intermittently. People inhibit congestion, jams, temporal uncertainties and health-threatening city environments, as a consequence of being encapsulated in a domestic, cocooned, moving capsule. (Urry, 2004, 28)

The spreading-out of important activities over a greater area means that social life involves connecting locations and places. Networks of locations and places with activities which take place there, are being created. Living means moving in networks (see Ohnmacht, 2009). Social life becomes intermingled with mobility. The urge to meet physically becomes essential in the different networks. Even in situations with permanent oral and written communication via mobile phone or Blackberry people want to meet physically. The car is then the third form of permanent communication.

Beck explained the loss of value of territoriality and proximity. Nowadays people with shared interests and lifestyles socialise more than those within a place of residence. People invest less in contacts within their vicinity and more in contacts with people with the same interests, hobbies or opinions. Travelling becomes essential to economic and social life and is no longer optional. Social duties and meeting meaningful people redefine travelling. Modern living, working and playing needs the capacity to move in dense geographical networks of social relations.

A culture of speed takes over older cultures. There is a need for permanent availability and constantly being on-line. The car facilitates individual timetables. To cite Urry, 'Automobility coerces people into an intense flexibility. It forces people to juggle tiny fragments of time so as to deal with the temporal and spatial constraints it generates. Automobility is a Frankenstein-created monster, extending the individual into realms of freedom and flexibility whereby inhabiting the car can be positively viewed, but also constraining car 'users' to live their lives in spatially stretched and time-compressed ways' (Urry, 2004, 28). And Paterson notices in *Automobile Politics* 'Automobility is the concrete articulation of liberal society's promise to its citizens that they can freely exercise everyday choices' (Paterson, 2006).

The operationalisation of mobility also changes. Mobility is no longer about going from A to B, but about integrating normal life with wished-for and needed social practices.[13] People are hurried, travel, are on-line to keep up the idea that they are behaving normally, according to expectations. Central in the Mobilities approach are the terms:

- spreading out of activities
- flexibility
- need to be mobile
- networks as an organisational principle
- lifestyles as a selection mechanism
- coupling of physical and virtual mobility

Urry noticed in *Inhabiting the Car* (2000b) that the car and the worldwide growth in car mobility was not paid much attention in the discussion about globalisation. However, the car organises modern societies. The journey with a car is seamless. Other transport modes are, compared to the seamless car, interpreted as inflexible and fragmented. Travelling by car you can avoid semi-public 'structural holes' like waiting at windy and unsafe stations and bus stops. The car structures time and public spaces and 'non-places of supermodernity'; locations and environments that exist only for the car and its users such as motels, parking lots, petrol stations and drive-in facilities.[14]

The Mobilities approach definitely has its own signature. Most of the literature is qualitative, with few data, almost literary work. Most articles are very descriptive, some with an excursion into philosophy. Most authors relate to and move onwards with views presented in other articles. The core authors are familiar with the work of Castells, Bauman and Beck. The Mobilities group fits perfectly into the Second Modernity.

---

13   An important Dutch study on this subject is by Harms: *Mobiel in de tijd* (2003).

14   There is some literature on the landscapes of car mobility: Normark (2006) writes about petrol stations, Merriman (2004) discusses the British M1, and the French philosopher Auge considers highway locations as non-places (1995).

The capacity to move: Motility

In the social sciences the term mobility is mostly associated with social mobility, climbing and descent on the societal ladder. In 2004 an old term from biology was coined for connecting social mobility and physical mobility. In this year the Swiss researchers Kaufmann and Joye published their article *Motility; Mobility as Capital*. They proposed a concept whereby physical mobility and social mobility are used as indicators for a more comprehensive form of mobility. They call this form motility, which stands for the capacity to move.

Motility includes the structural and cultural dimensions of movement and action, and has three elements:

1. access to forms of mobility
2. skills and competences to use this access
3. the competence to choose

Motility can be seen as a form of social capital, with access rights and competences.[15] In modern society, in the Second Modernity, the ability to move, with new concepts, new leaders, new organisational forms, has become an asset of its own.

In *Geographies of Social Networks; Meetings, Travel and Communication* by Larsen, Urry and Axhausen (2006b), motility is operationalised. The authors conclude that planning, in our time, is about organising simultaneous attendance. In earlier times meetings had to be planned long in advance and very carefully. With our online instruments, meetings have grown in flexibility. Real networkers know the conditions of modern network arrangements. The ability to move, store and disseminate information, to communicate just in time, and accurate in content is important. Appropriate meeting places are necessary, as are entrance to transport modes and a budget to ensure everything can be financed. When you do not succeed in arranging such conditions you can hardly be an important player in modern games.

There are some dilemmas in modern mobility and motility. The new mobility paradigm with its greater scale and its network pressures is not societal neutral. Richer and better-educated people do get more transport options. The carless households can face problems with their low motility capability. People who cannot properly use the new techniques and arrangements are at a loss and face social exclusion. I will focus on these themes in Chapter 6.

There is also a paradox with modern mobility. The traffic system is entering into fluidity. But the necessary infrastructure like roads, petrol stations and car parks are still long-standing elements. The fluid use of the system has a completely 'other' time scale than the building and maintenance of the infrastructure needed

---

15   For further operationalisation of the motility concept see Flamm and Kaufmann (2006).

for this system. The time scales differ; the pragmatism on the user's side is different from the long-term culture of planning and investing in infrastructure. Infrastructure cannot be made fluid.

There is also tension between the freedom and autonomy associated with the car and the need to have ever more rules for arranging the increasing traffic in a safe and appropriate way. Cars create new places, open up areas, but car can also destroy places and spaces, like children's playgrounds. The street is now for the car and no longer for neighbourhood living.

All these issues will have to be researched. A research programme for the social aspects of physical mobility seems appropriate. Sheller and Urry published the themes for such a research programme in *The New Mobilities Paradigm* (2006). They start by noting that many social scientists still use rather old-fashioned categories for their research, categories that are less useful in our complex realities in the Second Modernity. The authors consider six directions for further research.

The first is a revitalisation of the work of the German philosopher and sociologist Georg Simmel. In the 1930s Simmel presented an agenda for the study of movement. Second is the study of socio-technical systems as hybrids. Mobility needs people, networks, contacts, but also engines, power trains and information technology architecture. The third direction is the spatialisation of the social sciences. The connection between human geography and the social sciences has to be restored. The fourth direction is the study of the complex relationships between the traveller, the infrastructure and the transport modes. Fluid mobility needs inflexible infrastructure. As Sassen explains 'there is no linear increase in fluidity without extensive systems of immobility' (Sassen, 2002, 26). The fifth study area is the connection between social mobility and physical mobility, where nowadays different styles and cultures of research exist. Finally, the last study area is the analysis of complex systems. Here the focus should be on mobility in relation to the dynamism in societal arrangements.

2.9   Car mobility: Anxiety, uncertainty, expectations and chances

In Sections 2.4 and 2.5 I looked at two types of feelings which are structuring the emotional side of modern risk society. In Section 2.4 I focussed on anxiety and uncertainty, and in Section 2.5 on expectations and chances. Here, I focus on the relationship of these emotions with car mobility. Two themes are chosen; car mobility and children, and car mobility, leisure and shopping.

Car mobility and children

In modern societies children are responsible for a relatively large share of our mobility. For example, in the Netherlands children from 0–16 account for 11 per cent of the total mobility, on average 6,500 kilometres per child. Sixty-five per cent of this distance is travelled by car, 4,300 kilometres, almost 12 kilometres a day (Jeekel, 2011, 307). Note, that these figures do not include holiday mobility.

What are the motives? Of the 6,500 kilometres shopping takes 10 per cent, the school run 20 per cent, leisure 25 per cent, and visiting or staying with family or friends 30 per cent. For older children school mobility is higher and mobility related to visiting is lower than average.

The total child-related mobility will be higher. After escorting parents have to return. I estimate that around 15 per cent of total mobility and 12 per cent of car mobility is child-related.

In this Section I will focus on the school run and on escorting in general. There are great differences in the way children are taken to school between modern risk societies. To present some figures; in Germany in 2000 only 12 per cent of primary school children were taken to school by car (Limbourg, 2005). Most children walked. Somewhat higher figures are seen in Denmark; in 2000 nearly 23 per cent of 6–10 year-olds went to school by car, and 9 per cent of the 11–15-year-olds (Fotel and Thomsen, 2004, 538). For the Netherlands we have estimates available for 2010 (Jeekel, 2011, 320). Thirty-five per cent of the primary school children walk, 40 per cent cycle, and 25 per cent are taken by car.

In these three countries we note that children are, in the majority, taken to school till the age of nine, which means that on average 55 per cent of the children are accompanied by parents.

At the other end of the spectrum we have Australia and the United States. In Australia, in 2003, 65 per cent of children aged 5–9 were taken to school by car (van der Ploeg et al., 2008, 7, Table 1). In the United States, in 2005, 54 per cent of the children aged 4–15 were taken by car, 13 per cent walked, and 30 per cent took the school bus (McMillan, 2006, 77). The English figures are closer to Australia and the US. Mackett (2002) presented the 2002 figures; 41 per cent of the children aged 5–10 travelled to school by car, 51 per cent walked, and 7 per cent took the school bus. Cycling to school is not a feature of British culture.

Scotland and Belgium seem to be in-between. In Scotland in 2007 somewhat more than 28 per cent of the children between 4 and 11 years old went to school by car (Scottish Household Survey, 2008). For Flanders we have figures of 39 per cent of the 6–11-year-olds travelling to school by car (Zwerts and Werts, 2006).

With these figures three elements are noticeable. The first is that almost all the figures are relatively old. Early this century there must have been attention on the school run, in recent years aspects around the school run seem to have been taken for granted. The second element is the absence of standardised information; the focus in all countries differs, sometimes only the youngest children, sometimes all children, sometimes children and youth. The final, and most important, element is that figures differ so much. It looks like there is a North West European style of school run, with only 25 per cent car use, and an Anglo-Saxon style of school run, with at least 40 per cent but even 65 per cent car use. In Anglo-Saxon countries such as Australia, Canada, England, New Zealand and the United States it is clear that children aged nine and above still make accompanied school runs, where in

North Western Europe at age nine the majority of children are travelling to school unaccompanied.

In all countries car use for the school run has grown in the last decade. Mothers dominate this escorting to and from schools and working mothers, especially, take the car. Here we have figures from the United States (80 per cent of the mothers taking their children to school by car make a linked trip, mostly also to work, McDonald, 2005, 62) and Australia (in Melbourne, 61 per cent of car school runs are an element in linked trips, Morris et al., 2002, 6). From the literature a number of reasons arise for the high, or at least growing, amount of car use in the school run. First is the decline in the number of children, and hence of schools, in modern societies. The distance between home and school has grown. Second is the perceived lack of safety of the traffic along the route – parents are afraid of accidents and take their children to school by car, thereby increasing the chance of accidents for other children. Traffic safety of children went up in recent decades, however, this is an ambivalent success. As Huttenmoser explains in *Den Tanz met den Bandel* (2005) – the car has beaten the child. Children are not expected to use the streets for travelling or for play anymore.

Three other reasons plunge deeper into the culture of our modern risk societies. Many parents do not prioritize proximity when making their school choice. For middle-class parents school choice now is an important decision and part of a whole programme for education. Research in the Netherlands clarified that 40 per cent of the parents did not choose the nearest school, and accepted a longer school run (Jeekel, 2011, 313). Many parents are also anxious about 'Stranger, Danger'; they fear the public spaces their children have to cross. In 'Kids on the move' in Hutton and Peel (2003) O'Brien and Gilbert clarified that anxiety is behind many mobility decisions made by parents. Ridgewell et al. (2005, 8) and Morris et al. (2002, 7) also identified a related cultural element. In Anglo-Saxon countries it seems to be an aspect of good parenting to take your child by car, and not have him or her confronted by strangers. The last important reason is convenience. Cycling or walking takes more time, and time is scarce in our societies, where flexibility is such an important issue.

Looking more generally to escorting huge growth can be noted over the last two decades. In the Netherlands escorting is now responsible, for parents with children at home, for 32 per cent of the car journeys by women and 14 per cent of the car journeys by men (Jeekel, 2011, 327). Children, now also older children, are being taken by car to friends, hobbies, sports. Middle-class parents organise an education, inside and outside school for their children. Lareau studied this phenomenon in *Invisible Inequality; Social Class and Childrearing in Black Families and White Families* (2002) and comes to interesting conclusions. There is a difference between higher- and middle-class parents and parents of the lower social strata. Higher- and middle-class parents realise, in 'concerted cultivation', a whole programme for education, while lower-class parents leave their children free via what is called 'accomplishment of natural growth'. The mobility needs of higher- and middle-class parents are considerably greater; they have to travel to their network of friends, clubs, and sport areas.

This difference in attitude can also be seen in the school run. In the Netherlands there is a division among parents. Parents living further than 1,500 metres from a school, who are anxious and see their child as a very vulnerable human being, who do not choose the nearest school, and who have to travel to work afterwards, take their children by car to school. On the other hand, parents that live less than 1,000 metres from a school, look at figures and not at emotions, choose the nearest school, return home after the school run and almost never take their children to school by car (Jeekel, 2011, 319).

From the child perspective Karsten et al. (2001) see three types of children. There are the outside-oriented children. They play in public spaces, on the street, with relatively low supervision by their parents. These are mostly lower-class children. Then there is the 'back seat' generation. They are taken by car to friends, hobbies etc. Their lives focus in school and on a number of specific activity places.[16] They mostly do not know, and do not experience, the neighbourhood where they live. And finally, there is a group of children that seldom leaves home, being actively involved with computers.

This means that many children in modern societies are not growing up to be 'streetwise'. Middle-class parents have a low tolerance for danger and risks (see Durodie, 2005; and Furedi, 2001, 2002 and 2004). Until recently danger was seen as more-or-less being part of life. The modern middle classes now try to minimise risks and danger. As Thomson states in *How Times have Changed* 'children in the past are assumed to have capabilities that we now rarely think they have ... so fixated are we in going to give our children a long and happy childhood that we downplay their abilities and their resilience' (2009, 7).

Car mobility, shopping and leisure

Twenty years ago shopping in Europe was predominantly carried out in the neighbourhood and in the village or city centre. Over the last two decades many neighbourhood shops have disappeared and shopping centres and malls, located near to highways, appeared. The extent to which shopping has been relocated in the direction of highways and outlying shopping centres is different in Europe. France and Sweden are frontrunners in highway locations for shopping. But for example the Netherlands has only a few outlying shopping malls. In most countries furniture boulevards, do-it-yourself shops and garden centres now have locations outside the city centres.

---

16   Fotel and Thomsen (2004, 541) call this 'insularisation ... used to moving on the road only as a passenger between home and the scattered islands where he meets his friends to play with, rather than playing outdoors in the vicinity of the house'. Also, in German, '*verinselung*' (Funk, 2009).

Car mobility and outlying shopping centres are inseparable. Most customers get there by car. The French geographer Orfeuil (2004b) even mentions that in Ile de France, the region around Paris, 50 per cent of the commercial centres, with *hypermarchés* (huge supermarkets) and entertainment have no public transport whatsoever.

In *Consumption, Exclusion and Emotion; The Social Geographies of Shopping* (2001) Williams et al. discuss the new differences in shopping possibilities. With the split in shopping areas between the central city and the outlying shopping areas car users can easily visit both, whereas non-car users have to focus on the city centres. Especially in countries with many outlying shopping areas the city centres are dominated by 'fun' shopping, and are lacking the shops for daily essentials.[17]

Modern people like to shop at locations that are within easy reach, with everything under one roof and where they meet people like themselves. Maffesoli (cited by Williams et al.) calls this 'affective ambience'. Franzen mentions in *Retailing in the Swedish City: The Move to the Outskirts* (2004) that 70 per cent of Swedes use the car for their daily purchases. In outlying centres we talk about 90 per cent. Even when shops are close to their homes 47 per cent of Swedish consumers would still prefer to use the car. Most frequent consumers are middle-class families. The cheapest supermarkets in Sweden are to be found in outlying areas, unreachable for the poorest households. This is the 'poor pay more syndrome' (Franzen, 2004, 100). Related to this syndrome is the existence, in North America, of so-called 'food deserts'. In *Mapping the Evolution of Food Deserts in a Canadian City* (2008) Larsen and Gilliland explain that the poor households in a city can only buy food in expensive, small delicatessen shops, since there are no supermarkets left in the cities.

Mobility for leisure is, in most modern risk societies, measured in passenger kilometres, the most important category of mobility. However, leisure mobility is a less standardised category than shopping mobility or work mobility. From a perspective on car mobility leisure can be split into four groupings:

1. 'staying at home' (the garden, computer use, hobbies, meeting friends),
2. 'near the house' (a short walk, a short cycling tour, going to the library, sports),
3. 'at greater distance' (longer cycling tour, walking trip, fun sports, going to amusement parks)
4. 'really far away'(camping, holidays).

A number of activities can fit in two groupings. In general, with growing distances car dependence for leisure activities rises. Car dependence is higher when luggage is involved, and car dependence is higher in the evenings, because of the scheduling of public transport and anxiety about being out late in public spaces.

17　For example, some years ago we could not find milk in the centre of Annecy, it could be obtained only from the *hypermarché* near the highway.

A leisure activity near home, in the daytime, with no luggage is always less car dependent then a leisure activity at a greater distance, in the evening and with luggage. This scheme becomes more difficult with activities in two criteria; is the car dependence of a leisure trip near home, with much luggage, in the day higher or lower than a leisure trip further away with no luggage, in the evening? Compared to work-related mobility there are very few data on leisure mobility, but intuitively I will state that probably the order is distance–luggage–moment.

Leisure mobility has grown, but people in modern risk societies have less free time than 15 years ago. In *Vrije tijd in een tijdperk van overvloed* (2003) Mommaas identifies the paradox that people with lower spending budgets have more free time than people with higher spending budgets. The higher spenders, with less free time, need speedy travel to reach leisure destinations. Some segregation in leisure arises; hanging around in cities for the lower incomes, and for the singles, and travelling to leisure locations in suburban and rural areas for middle-class families. In general the free time is more consumer-oriented, more dynamic, and more experience- and adventure-oriented than in the past.

Table 2.4   Overview on anxieties, risks, chances and expectations

	Anxieties	**Risks**	**Chances**	**Expectations**
Children	Unsafe traffic	School environments difficult to read with many traffic modes at once	Car is convenient	Good motherhood means avoiding difficult situations for children
	'Stranger Danger'	Children do not get traffic education	Car makes it possible to choose a school at greater distance	Children's lives are structured and organised by middle-class parents
		Children miss experiences in their own neighbourhood	Patterns of mutual help between school parents arise	
Leisure	Aggression in the evening around public spaces	Shopping more expensive for lower incomes	All potential leisure spots can be reached	Funsports: everything that has been developed should be tried

2.10   The risk society and car mobility: Some first conclusions

After this kaleidoscopic picture on car mobility related to core elements of modern risk societies I will now return to the list of characteristics, presented in Section 2.6. Some characteristics have a direct relationship with the appearances of car mobility, with others there is no relationship. My list looks as follows.

Table 2.5   Characteristics of risk society in relation to car mobility

Hardly a relationship	Indirect relationship	Direct relationship
Abstract systems	Wish for simplicity and authenticity; to be seen in leisure, with funsports	Implicit expectations in the social sphere; Good motherhood demands travelling around by car, in bad weather you bring your child by car
Moving forward without direction	Expectations and wishes; People expect other people to have a car, time schedules and activities are planned on car availability	Fears, feeling vulnerable, seeking security; Lack of safety in car traffic and Stranger Danger in public spaces have to be fought, the car is a helper.
Scepticism about science and technology	Differentiation in lifestyles; not easy to take a passenger with you	Flexibility as a necessity; Cars useful by criss-cross patterns, with luggage, and with many short trips in a chain, in a fixed time schedule
	Leading roles for the media; important in generating anxiety, with car use as a mitigator	Everything on appointment; Car is better for last minute travel
	Active leisure as an asset; Broader and more active spectrum create greater distances to reach leisure spots	Geographical spreading out of activities; e.g. peri-urbanisation and highway locations

The picture is diverse. Not all characteristics of modern risk societies are related to car mobility. However it is clear that at least:

- implicit expectations in the social sphere,
- fears, feeling vulnerable, seeking security,
- flexibility as a necessary asset,
- everything on appointment, and
- geographical spreading out of activities,

have a clear relationship with the rise of frequent car use.

Chapter 3
Attitudes and Motives for Car Use

In Chapter 2 I looked at car use related to the development of the modern risk society. Car use is however not completely dominated by the characteristics and development of the modern risk societies. Car use is also a result of many individual attitudes and decisions. These attitudes and decisions are elements in societal contexts that stimulate car use, but they remain individual choices.

Attitudes describe basic orientations towards car use. In Section 3.1 I will introduce a spectrum of attitudes related to cars and car use. From Section 3.2 on I will focus on individual motives. In the literature many motives can be found and each author on motives for car use needs to decide some form of classification. With, as a basis, many articles (of which the most relevant are; Maxwell (2001), Steg (2001), Hagman (2004), Steg and Gifford (2005), Gatersleben (2007), and recently Schwanen and Lucas (2011)) I have chosen five motives:

1. convenience
2. flexibility
3. protection
4. freedom
5. habit

In Sections 3.2 to 3.6 I will look for situations, activities and expressions that can be considered characteristic for these motives. I will present the relevant aspects related to the motive and the core literature on the motive and I opt for an active use of the language. The reader should be able to position him- or herself in the motives presented.

3.1 Attitudes of car drivers

Mostly the standard statistical categories were followed; age, income, gender, education. Most social scientists typologise the population by these categories. However, Beck (1992) calls these categorisations of data 'zombie categories', and considers them old-fashioned, and not able to clarify our complex risk societies. In his eyes the world as it develops can no longer be understood by these categories. Our society is now so open, transparent, individualised and complex that people with the same statistical profile (on income, gender, age, education) can define their chances, possibilities and problems completely differently, and hence can lead completely different lives. Beck gets support from marketing experts, clarifying that the population can now better be distinguished by attitudes and lifestyles than by the normal statistical categories. I will present in this section three classifications starting from these notions.

The first classification comes from the Dutch project *Stedenbaan*, and has been elaborated by Verburg, Schwanen and Dijst (Verburg et al., 2005). Journey statistics of 29,000 persons have been used, and three types of travellers were identified: car users, public transport users and walkers or cyclists. With cluster analyses the three types of travellers have been classified in so called mobility styles. For the car users there are six mobility types:

1. the caretaker; car users travelling almost exclusively to care for kids, family, friends, 25 per cent of the car users, mostly women
2. the visitor; has much free time, will not drive at congestion periods, 16 per cent
3. the relaxed; car use for leisure, often in the weekends, 19 per cent
4. the traditional commuter; drives to and from work, and nothing else, 30 per cent, mostly men
5. the student; uses the car for travelling to education, 2.5 per cent
6. the business professional, much on the road, 7.5 per cent, mostly men.

In the view of the authors, each style needs a specific approach.

A second classification has been developed in Germany, by Gotz, who chooses in *Mobilitatleitbider und Verkehrsverhalten* (1999) a classification of attitude and lifestyles. He identifies four dimensions:

1. a recognised need for security and protection against threats
2. a drive for adventure, variety and risk
3. the social status as shown by your car
4. the orientation towards nature.

Gotz looks at consistent lifestyles, which he exemplifies as showing orientation towards nature, disinterest towards social positioning, no adventure seeking, and a need for security. A car user with this attitude differs completely from an adventure seeker, interested in social positioning. Gotz identifies in his research that among the car users in general no gap between sustainable objectives and non-sustainable driving practices can be analysed. The real situation is that distinguishable groups of car users exist which hold completely different positions towards sustainability, and that these groups are rather consistent in their values and attitudes. Gotz considers it impossible to work with statistical categories, but recommends a differentiation between distinguishable groups of car users.

I will present a third classification in somewhat more detail. The Dutch marketing advisor Motivaction classifies the Dutch population in eight mentality groups (Motivaction, 2002). This classification is based on the assumption that people with the same socio-economic–demographic profile can have completely different lifestyles and consumption patterns. Motivaction argues that who you are, what you think, what you appreciate and what you do is more dependent

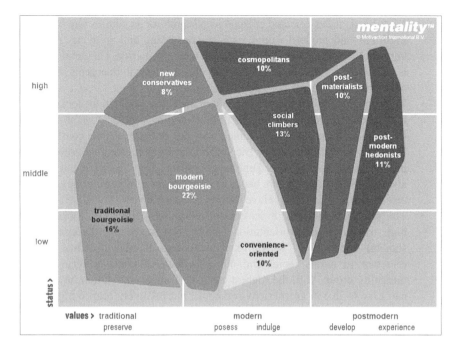

Figure 3.1   Mentality milieus in the Netherlands
Source: Motivaction, (2002).

on personal preferences, norms and values than on statistical characteristics. In Figure 3.1 the classification in mentality groups is presented.

On the Y-axis the socio-economic status is presented; low, middle, high. The X-axis has a new variable; a value orientation. Three value orientations are identified, from left to right; traditional (preserve, conserve and maintain), modern (possess material things and indulge yourself) and post-modern (develop, learn and experience). Motivaction has created a huge database, and works with eight mentality groups that I will shortly introduce, with their focus on car mobility.

- The traditional bourgeois: 16 per cent of the Dutch population. Social class: low and middle, traditional orientation with focus on preserve and maintain. Can be seen as: the status-oriented dutiful citizens who keep traditions and material ownership alive. Family and friends live nearby, work is also at a short distance. This is the least mobile mentality group, is mostly rural and buys mostly new cars. They drive quietly, and are not very car dependent; they like travelling by car, but they certainly see and use alternatives.
- The modern bourgeois; 22 per cent of the Dutch population. Social class: middle, modern orientation with a focus on possession. Can be seen as;

conformist, status-oriented citizens who search for an equilibrium between traditions and modern values such as consuming and enjoying. Are not very mobile. Their car is a symbol of their prosperity. The action space is relatively small, but greater than the traditional bourgeois and the convenience oriented.

- The convenience oriented: 10 per cent of the Dutch population; social class: low, modern orientation with a focus on indulging yourself. Can be seen as: the impulsive and passive consumer, who strives for a carefree, comfortable life. They would like to have friends and family in the neighbourhood, but this is not always the case. Small action space. They like cars, but cannot always afford a car, and are hence as a group not very car mobile. Are not oriented to sustainability. Have more risk-taking driving styles.
- The new conservatives; 8 per cent of the Dutch population; social class: high, traditional value orientation. Can be seen as: the liberal-conservative societal upper layer, which likes technological advancements, but is reluctant to adopt social and cultural modernisation. A very mobile mentality group, with the highest car use of all groups. Often buy a new car. Like to live in greener environments, but relatively near to highways. Friends and family are dispersed, great action space.
- The cosmopolitans: 10 per cent of the Dutch population; social status: high, modern value orientation with a focus on indulging oneself. Can be seen as: the open and critical modern citizens who live in and with the world. They combine modern and post-modern values; they like to learn, to develop themselves, and love societal success. Are rather materialistic. Live more often in cities, are very mobile, but not very car oriented. Make the most air miles. Friends and family everywhere.
- The social climbers: 13 per cent of the Dutch population; social class: middle, modern value orientation on indulging oneself. Can be seen as: career oriented individualists with a fascination for social status, new technologies, adventures and risks. Rather mobile, family and friends live at a distance. Like to live near highways. Car is a status symbol. Little use of public transport, very little walking and cycling.
- The post materialists: 10 per cent of the Dutch population; social class: high, postmodern value orientation. Can be seen as: societal focussed idealists who want to develop themselves, and who will aim at sustainability and social equity and welfare. Family and friends live at a distance. A highly mobile group, the most mobile of all groups, and second in car use! Very busy with combining tasks. Tries to use public transport, but often fails.
- The postmodern hedonists: 11 per cent of the Dutch population, social class: middle, postmodern value orientation, with a focus on experiencing. Can be seen as: the pioneers of the spectator culture, experimental, and breaking with moral and social conventions. Live mostly in cities, like to spend their evenings outside home with friends. Are mobile, car not very important, highest use of public transport of all groups.

The great differences in these lifestyles, related to car mobility seem to focus on:

- the action space
- the importance of the car as a status symbol
- the importance of living near highways.

We can identify:

- the wish to live near the highway,
- the car as status symbol, and
- great action space,

as indicators for a strong reliance on the car. The new conservatives and the social climbers are very car reliant, and the traditional bourgeois, the post-hedonists and part of the modern bourgeois are not car reliant. The other three mentality groups are in between, with a special position for the post materialists. They are ambivalent in their orientation, as they see all the disadvantages of car use in relation to sustainability, but need car mobility in their busy lives. The convenience-oriented also show some ambivalence; they like driving, but cannot often afford the purchase.

Motivaction (2002) identified four mobility styles, related to their mentality groups:

1. The *local traveller (Lokaal)*; he or she does not travel much. Here we see the traditional bourgeois and modern bourgeois, 30 per cent of the Dutch population.
2. The *convenience traveller (Noodzakelijk en Gemak)*. Here we see the convenience oriented and also the post hedonists; they do not travel that much by car, 20 per cent of the Dutch population.
3. The *private and status oriented traveller (Privé en Status)*; this is by far the most car reliant group, the modern conservatives, the smaller part of the modern bourgeois and the social climbers, 30 per cent of the Dutch population.
4. The *responsible traveller (Openbaar en Verantwoordelijk)*; they drive, are very mobile, also by car, but feel responsible with regard to sustainability.

From this classification two clear groups and two ambivalent groups arise. Local travellers are not very car-oriented and have a small action space. Their counterparts are the private and status oriented travellers. They are very car-oriented with a large action space. The ambivalence is with the two other groups. Convenience travellers are car-oriented, but have fewer cars than expected and responsible travellers are not very car oriented, but use their cars a lot.

This analysis shows that 50 per cent of the population will define themselves as car-oriented; the private and status travellers and the convenience travellers.

The other 50 per cent do not define themselves as car-oriented; the local travellers and the responsible travellers. However, these two types are in practice sometimes rather car oriented. The core of the non-car orientation in vision and mentality is with the post-materialists, the cosmopolitans, and the traditional bourgeois, comprising 36 per cent of the Dutch population.[1]

Individual motives

I will now focus on the individual motives for car use. An interesting reader on emotions related to car use, *Automotive Emotions*, was published in 2011. In 13 articles a broad spectrum of emotions related to car mobility was presented by the editors Lucas, Blumenberg and Weinberger. In this book the focus is broader – not on car dependence and on frequent car use, but on all aspects of car use in the broadest sense – and on the other hand smaller in scope than in my book. The scope of the book is restrained to understanding emotions. No analysis of the impacts of choices for car mobility on the functioning of modern western society is offered. Different theories to look at automotive emotions are being presented, driving forces are introduced, and half of the reader is about the choices of different societal groups like dual wage-earners in the Netherlands, older US drivers, low-income Australians, poor British car users and migrants to the US. The editors conclude from their presented material that automobility is extremely complex, and that 'one size fits all' – policy responses will be inappropriate (Lucas, Blumenberg, Weinberger, 2011, 282).

As mentioned, in this chapter I will focus on five composed motives for frequent car use.

3.2   Convenience

Convenience stands for; 'everything used to diminish frustration, time spend, or energy spend'. Convenience is a relative term. Convenience is an important motive in our Western European societies. Galbraith states in his book *The Culture of Contentment* that the majority of the population of the richer countries is primarily interested in a safe, comfortable and easy existence. To cite; 'the fortunate and the favoured, it is more than evident, do not contemplate and response to their longer-term well-being. Rather, they respond, and powerfully, to immediate comfort and contentment.' (Galbraith, 1992, 6). In the modern western risk societies personal self-discipline and restraint are certainly not core values.

People associate the car, more than other transport modes, with convenience. A few arguments need to be mentioned here.

With a car you can leave wherever you are when it suits you. This is also the case with walking or cycling, but certainly not with public transport. You can

---

1   Within the post-hedonists are also people with a non-car orientation.

also depart and arrive at all possible times of the day. And mostly, depending on parking regimes, you are able to end your journey in the vicinity nearest your destination. Not being dependent on institutional time tables but being able to follow your own, and not being dependent on extra pre- and post-transport periods are convenient advantages of the car.

With a car you are not confronted with the weather conditions. Cold, rain, wind can be 'switched off' in a car. For several decades we have been trying to lessen external influences on our bodies. We do not want physical proximity of other people, and we will not be cold, not be sweating, not be wet. At least, we do only want to feel these external influences in controlled settings. Sweating is for exercise, or on Sundays on your sports' bike, the wind is only allowed to touch you when sailing or paragliding, and physical proximity is only acceptable in the presence of friends and partners. Outside these controlled situations we do not want the weather influences or physical proximity to affect us. The car makes this possible. Cars have developed with regard to comfort in recent decades; individual microclimates and individual listening environments can be created in your car.

With your car you do not have to travel with heavy luggage across public spaces. The car takes your burden. Without a car you mostly have to pay for the delivery of goods.

The car also saves time and frustration. The extra energy generated by these savings can be used for other objectives. In one of the scarce studies on convenience (Shove, 2002a) is signalled that the convenience of the car is mostly appreciated when coordination problems exist, and when individuals are obliged to follow time schedules and specific routes to achieve all their daily objectives. In these situations driving does not mean going somewhere. Driving then creates the possibility of arriving and departing at exactly the right time to achieve the interlinked objectives.

As people grow accustomed to car use, they tend to become more averse to physical effects, or efforts outside controlled spheres. In 2005 Garling and Loukoupoulus published their article *Are Car Users too Lazy to Walk?* They came to the conclusion that convenience mostly becomes a habit. Habitual drivers do not want to walk, even for short distances.[2] However, a lot of short journeys by car are environmentally more damaging than fewer longer distances as most emissions are produced at the start of the car journey.

People grow accustomed to convenience. Comfort standards in cars rise over time, and car manufacturers produce global cars for global markets, so; cars have more or less the same standards all over the world.

Rising standards of comfort (micro-climate, gadgets, air-conditioning, radio, audio, DVD etc.) lead to higher energy use (see Heywood, 2008; Cuenot, 2007; Mol, Lavrysen and Vlassenroot, 2006).

---

2   In *Driving through Car Geographies* (2006), Sarmento gave an example of Portuguese students at the University of Minho. They preferred to take the car instead of walking 500 metres from their dormitories to the lecture buildings.

Rising convenience levels, mostly through the introduction of higher codes and standards by the producers of consumer products lead to more energy use, and to less sustainability. Research shows that the levels of expectations levels within households rise related to these higher standards (Chappels and Shove, 2004). For example, in the UK households now set higher ambient temperatures, and the practice of machine-washing clothes has intensified.

Measuring convenience is not easy. Convenience in most questionnaires includes the flexibility that can be reached by cars. I introduce the motive flexibility as a separate motive in 3.3. Here I will operationalise convenience in four aspects;

1. being able to create a micro-climate in the car
2. being able to eliminate weather influences
3. the possibility of immediate use
4. the use of the car for luggage.

The first two aspects are about convenience in the sense of comfort, the last two make life easier.

In the scientific literature very little is published about convenience.[3] There are only a few guiding studies.[4] Chappels and Shove concentrate in their study *Comfort: A Review of Philosophies and Paradigms (2004)* on the regulation of climate and temperature in houses. They signal that the growth in convenience mostly does not find its source in consumers' wishes, but in the work of pioneer technicians, who develop higher convenience standards. Chappels and Shove identify a great variety in levels of comfort and in expectations of comfort between households and countries.

What is seen as comfortable shifts over time. To give an example: around 1960 the inhabitants of Chicago moved their sleeping bags to the garden during extremely hot weather. Now they set the thermostat of the air-conditioning to a cooler level, using a lot more energy. In this pioneering study the authors were sceptical about the possibility of creating modern comfort expectations that could cope with sustainability objectives.

The aspect of sustainability is also an issue in *Revisiting the Construct of Convenience in Consumer Research: The Paradox of Convenience Consumption* (2004). The authors, Carrigan and Szmigin, notice that convenience products can save large amounts of time. Consumers see the tension between the care or precision needed for a task (e.g. cooking) and the use of convenience products

---

3   Almost nothing is published about transporting luggage as an aspect of convenience. The only article known is Messerl and Troch (2002) regarding car use, luggage and winter sports.

4   This is perhaps a little strange; the constant rise in convenience levels for all sort of activities and products scan be seen as an important reason for non-sustainability. On this theme in a broader perspective, see the work of Jackson (2004a), on sustainable consumption.

(take-away meal). Consumers will mostly try to mitigate a complete use of convenience products. Annexed to this behaviour they introduce the paradox of convenience, using convenience products makes life somewhat easier, but the standard of the activity for which the convenience product is a substitute rises, and so new convenience has to be realised. A good example is the practice of machine-washing clothes; more convenient, but also carried out far more frequently than a few decades ago.

In *The Consumption of our Discontent (2005)* Sanne introduces a political perspective on convenience. Politicians seem to be able to combine a discourse on frugality with a discourse on stimulating trade and industry. In her view this last discourse will lead to higher consumption levels. She notes that consumers differentiate between very useful consumer products and less necessary consumer products. She cites a study from Sweden from 1995 in which a telephone, a freezer, healthy living and a television were seen as more necessary than a car, but where cars were seen as more necessary than a refrigerator, a microwave and holidays.

Do car users take measures to mitigate their reliance on convenience, as we saw with convenience foods? It looks as if it is felt necessary to compensate for too much convenience. On the one hand; the car is not used for every trip, at least not with in the majority of the population. On the other hand; jogging, fitness, hard cycling, can be seen a compensating instruments for too much convenience elsewhere.[5]

Finally, convenience serves mankind, allows extra activity, creates comfort. But convenience can, indirectly, also lead to stress. Because it is easy to use the car for journeys, *you, and only you*, will become responsible for being somewhere on time. You cannot make a service provider, a public transport provider, responsible. Only you are responsible for getting your children to their hobby on time, or for collecting them from the kindergarten. In modern risk societies this responsibility creates stress, because although the success of your trip depends on other drivers (think for example about congestion), there is no way to make them co-responsible for your arriving on time.

3.3 Flexibility

Flexibility as a motive includes manoeuvrability, agility and speed. It can be identified as; 'useful and most usable for the total spectrum of daily travel objectives, without much investment in advance'. The car is flexibly usable, is agile and fast and seems to be the only mode of transport able to connect swiftly

---

5 Use of these instruments differs from cooking more complex meals. This is a real alternative and a substitute for convenience food. Jogging, hard cycling etc. cannot be seen as an alternative to convenient driving. Jogging etc. is not integrated in daily journey patterns. There are separated mobility worlds, one for real life (the car) and one for leisure (jogging etc.).

the different locations where important activities for members of the household are taking place.

Our modern society is, for households, especially those with schoolchildren, and for singles, a society where combining tasks is standard. *On each single day lots of things and activities have to be arranged.* The children must go to school, you would like to visit your elderly, lonely aunt, you work part-time, the dinner must be served, you have some shopping to do, and the plumber is coming, but cannot say when exactly he will be near your home.

This is written from a woman's perspective. Women are, in the modern risk society, the great multi-taskers.[6] A similar example can be used for men, and especially for fathers, on Saturdays and working singles have to arrange all the essentials of life themselves and sometimes make a higher number of journeys on weekdays than working mothers (MON, 2007).

In modern risk societies task combining has grown more difficult. Necessary activities for households such as shopping, escorting children, and helping friends and family with health problems, are now located at a greater distance from each other. Shops, sporting facilities, schools and offices are now situated at a greater distance from homes. Walking or cycling to these locations is sometimes difficult.

In *Unvermeidlich wie die Kautzen in Athen ; Anmerkungen zur Soziologie des Automobils* (2003) Rammler presents an analysis of processes on a greater scale that are basic for the need for permanent car use. Private mobility creates spatially spread location patterns. Important for these patterns is the combination house/ car. When people take a new job further away, they mostly do not move house, but travel longer distances. After some time the adults in the household, both working, can live at relatively great distances from their work locations. Especially when the two work locations are in different direction they will stay in the same house, thus increasing the number of kilometres travelled. In principle this situation shows the relatively low price of car driving, in monetary terms. Rammler states; 'Woheigentum hemmt residenzielle Mobilitat unter gleichzeitiger erstatzweiser Ausweitung der Fahrleistungenb in der zirkularen Mobilitat' (Rammler, 2003, 11; 'house ownership inhibits housing mobility and creates extra rushing around').

Daily life has also grown more complex and more restless. Growing prosperity made more explicit choices possible. We direct our activities – send children to specific clubs and hobbies and, more important, we often choose not to use the nearest service or facility, but the ones we like best. This can be seen with schools, shops, but also with friends. Previously, in the 'housewives' period, women had the whole day to do the necessary daily activities, now they have to do more in less time.

Daily activities have to be arranged in rather tight timeframes. Opening hours, most shopping hours, hours at which technical services such as plumbers, can be assessed, school hours and working hours in most modern risk societies still follow

---

6   To present a statistical example; in the Netherlands women between 30 and 50 years make 20 per cent more trips than men of the same age, but travel only 65 per cent of the men's distance (MON 2007, tables 8.10 and 8.11). Women have their bias on many short trips.

a 'nine to six' pattern (Breedveld, 1999). Many activities in a tight time frame make a system of time planning essential, with a number of 'just in time' situations. This creates feelings of time pressure and sometimes stress or hurriedness. To cite Knulst, a Dutch researcher in the aftermath of this development, in 1984: 'persons working within normal working hours have a 10 to 15 per cent chance of finding a service they need open' (Knulst, 1984). The Dutch Social Research Institute (SCP) concluded 26 years later; 'the situation where more hours are being worked, while the opening hours of services have not changed significantly leads to the suspicion that the possibility finding a service open probably has diminished' (SCP, 2010, 110).

Our modern risk societies are characterised by *a battle around time*. As long as this battle is not made explicit, and not fought, feelings of time pressure, hurriedness and stress will probably grow. The Dutch Social Research Institute (SCP) presented in 2004 a report called *De Veeleisende Samenleving*. This report contains an essay on stress and stress feelings in our times. Their principle is:

> *De moderne maatschappij is een veeleisende maatschappij. Nieuwe mogelijkheden worden vooral benut om nog meer te kunnen realiseren, niet om de inspanning te verminderen en meer tijd over te houden. Sneller vervoer leidt tot grotere reisafstanden, hogere productiviteit tot meer productie, meer inkomen tot meer consumptie, emancipatie van de vrouw tot meer tweeverdieners en hogere hypotheken. 'Meer en snel' genieten in de huidige maatschappelijke constellatie een hogere status dan 'minder en rustig'* (SCP,2004,36).

> Our modern society is a demanding society. New possibilities are used to realise more, and not to reduce efforts and strains, thus realising more leisure and free time. Faster mobility leads to greater travel distances, to higher productivity, to more production, the higher incomes, to higher consumption, to women emancipation, to more task combiners and to higher mortgages. 'More and fast' is in our societies more important than 'less and relaxed'. (author's translation)

The price for this choice is high; a relative high degree of mental fatigue. But from the perspective of the SCP *'kennelijk onvermijdelijk neveneffect van een maatschappelijke versnelling en een kennelijk voor wenselijk gehouden individualisering'* [this seems to be an inevitable side effect of an acceleration society and an accepted and wished individualisation – author's translation] (SCP, 2004, 36). Our existence is time-intensive and with this time – intensity comes, according to Breedveld who studies school hours, childcare and the related responsibilities; *'één kink in de kabel, één trein die te laat is, één vergadering die uitloopt, of één dag dat de kinderen niet naar school hoeven, en er ontstaan problemen'* [one hitch, one train too late, one meeting that ends later, or one day that children are unexpected free from school, and problems arise – author's translation] (Breedveld, 1999, 22).

Southerton presents in *Squeezing Time: Allocation Practices, Coordinating Networks and Scheduling Society* (2003) a study on hurriedness. He interviewed

20 households in a suburb of Bristol. All respondents felt sometimes, and a few very often, hurried and nervous. The need to have a dual, or at least a one-and-a-half income in order to be able to live what is considered a normal middle-class life in our societies leads to less time for daily household activities. Whether hurriedness existed 'depended on individual capacities to schedule their practices in line with the shared socio-temporal constraints found in their contexts of network interaction' (Southerton, 2003, 17). Hurriedness depends on context. When a few of the planned activities are not taking place as scheduled a relaxed morning can easily change into a hurried day. From his interviews Southerton concluded that hurriedness is increasing.

There is another reason for this feeling of time pressure. Extra speed to overcome distance is not used to diminish time pressure. It is 'consumed' through shortening the time periods between the activities, allowing car divers to carry out more activities in the same time frame. Greater flexibility leads to greater flexibility. To cite the Dutch philosopher Achterhuis, 'people buy time with speed, and this process continues and continues' (Achterhuis, 1998). In this circumstances, stress and time pressure remain, but this can be put in perspective; 'not only is time pressure normal, it is also socially acceptable and to a degree status-giving ... time pressure becomes "a contemporary myth"' (Hjorthol, 2005, 7).

Modern households need a transport mode that gives them the ability to combine many activities and appointments at different locations in a tight time schedule. The car seems to be the only transport mode that, at least for very many households, fits into these demands. In *Rushing Around: Coordination, Mobility and Inequality* (2002b) Shove analyses what is necessary to allow the social practices defined as normal in our societies take place. She defines a practice as; a routine like way in which people travel, use products, in which developments are framed, and in which the world is understood. Mobility systems facilitate all the practices that can take place, but mobility systems do also change these practices. The task for mobility changes through time 'mobility is not about getting from A to B ... but instead about integrating everyday life and the activities required of 'normal" practice. And, 'people are rushing around in order to preserve the sense that they are behaving in normal and ordinary ways' (Shove, 2002b, 9).

Each time this deals with 'spacing and timing', planning and organising of different activities and practices in space and time. Each human being in our time and society has to follow his or her own spacing and timing programme. In Shove's view the car is best-equipped to help with this 'rushing around'. Exclusion can be a problem for people obliged to participate in these practices without the possession of individual and collective instruments to follow the aforementioned difficult and tight time schedules. Shove concludes that a more collective programmed society will create less social exclusion, and she asks what in our time is considered to be 'an effective membership of our "normal" society'.

Cars organise the traffic between the nodes in the 'interest and attention networks' of modern households. Much of what nowadays seems to be a normal life with friends at greater distances, with many experiences in different areas of life at several locations, is rather difficult without a car, unless one is rich or very creative. Often, modern people have the idea that they do not have a realistic choice not to use their cars. Households often rely on the travelling-round mothers with time pressures. As the German researcher Rammler says, the car grows into 'ein Grundausstattung eines volwertigen Gesellschaftsmitglied' (Rammler, 2003, 5, ['the car grows into a basic element for full and able membership of modern society']).

Not having, or better, not using a car can means – unless a large budget to buy alternatives is available – that fewer activities can be undertaken or facilitated in a day. Possibilities and chances can be missed. In many countries the attention to this element is rather marginal.[7] It appears that most countries hypothesise that most carless households will be 'time rich', and that social networks around these carless households will still function.

For the 'time poorer' households car use seems necessary. In *Running Around in Circles* (2003), Skinner analyses the need for mothers to manage a number of deadlines each day. On time for school, not too early leaving work, on time back at school, on time to the hobby, and on time picking up from hobby. Managing the coordination points, the moments of change to another activity is essential. Keys to successful management of deadlines are a short distance between work, school and care, flexible working hours, help from family and friends, and having disposal over fast transport.[8]

Dobbs describes in *Wedded to the Car; Employment and the Importance of Private Transport* (2005) some reasons why households with access to many public transport facilities still use their cars for most journeys. Public transport does not take them exactly where they have to be, and households are very critical about the inability of public transport to make chain trip patterns. Public transport seems to be stuck in a 'nine to six' – society. Women are more active on the labour market when they have a car at their disposal.

In *Famille et temps: Etat de l'art et tour d'horizon des innovations* (2002) the Swiss researchers Kaufmann and Flamm present a European perspective on this theme. Looking at organising life with children they see three models.

The first is the traditional model; the majority of women stay at home to take care of the children. This is the case in Italy, Greece, Spain and Ireland.

---

7 Note that in the UK for some years a Social Exclusion Unit has existed within the Ministry of Transport. In France there are *Bureau de Temps*, to look at the (uneven) distribution of available time and time pressures.

8 Craig (2005) signals in *How Do They Do It? A time-diary analysis of how working mothers find time for their kids*, that full-time or nearly full-time working mothers with children in full day nurseries spend almost the same amount of time with their children than other mothers. The car is necessary to make this possible. A view arises; 'women rushing from work to pick up their children from day care, cooking, bathing and feeding and talking to and playing with and reading to their children, and cuddling them to sleep, before dropping exhausted into bed themselves and beginning it all again the next day '(Craig, 2005, 16).

On the other end of the spectrum are countries where most women work more-or-less full-time. This means leaving their children in controlled environments for the whole day. This is the case in France, and mostly in Scandinavia.

The third group is an intermediate group. Women mostly work part-time, and children are, part of the time, raised within the family as in Italy etc. This is more-or-less the situation in Great Britain, in Germany, in Austria and certainly in the Netherlands.

Countries supply different arrangements. The authors show that, for example in France, part-time childcare is mostly not available. The three models have direct consequences for car mobility. The relatively relaxed situation in Italy etc. has in the intermediate countries – combining work and child care on the same day – become a rushing around pattern. In the countries with full employment for women the stress is concentrated on the two moments of leave and pick up. Time in the car becomes quality time; *'pour les meres actives a plein temps, les deplacements avec leurs enfants sont meme consideres comme un maniere d' avoir un contact privilegie avec ses enfants au quotidian'* (Predali, 2002); ['for women that work all day, the journeys with their children are seen as a way to have privileged contacts with their children'].

Giger shows, in a study in the Lausanne region called *Une perspective de genre sur la mobilite quotidienne* (2008, 46), a geographical division. In areas with high population densities men take their responsibility for child care far more seriously than in areas with low population densities. In these areas, the peri-urban areas that I described in Chapter 2, mothers do all the work related to children and spend less time working outside the home.

Wajcman works in *Life in the Fast Lane* (2008) from an acceleration perspective.[9] In her eyes there is no general time shortage. In caring for children the clock time is less important because, 'the direct activities parents engage in with children consume far less time than the responsibility for overseeing them' (Wajcman, 2008, 65). Wajcman also has an explanation for the greater feelings of stress among women; 'when faced with a finite 24 hours in the day, mothers identify the use of time by multi-tasking and juggling short spells of caring activities with equally short spells of leisure and care '(Wajcman, 2008, 73). Leisure comes, for women, not as a block, but as short spells.

Also human geographers have thought about patterns in time and space. In 1970 the Swedish geographer Hagerstrand introduced his time-geography. In essence this geography focussed on arranging activities in time and space combined. He introduced terms like time-space tracks, time-space restrictions, and projects. A track is the result of the interaction of projects and restrictions. When this theory was introduced it was seen as interesting but unworkable. His theory is now revaluated, for example by Schwanen. In *Matter(s) of Interest; Artefacts, Spacing and Timing* (2007a) he relates the theory to modern cares around 'spacing

---

9   The theorist of this perspective is the German sociologist Rosa. See his *Beschleunigung* (2005).

and timing' of double-earner households. He looks at daily nurseries and shows that even searching for children's toys can make a normal day into a hurried day.

In *City Time; Managing the Infrastructure of Everyday Life* (2004) Jarvis focuses on the political and societal context of time scarcity and hurriedness. She argues that time scarcity is too often framed as an individual problem, and an individual question but time scarcity also has power elements and inequalities. Richer households can more easily mitigate time scarcity than poorer households and responsibilities for service providing previously taken by governments are now, in the neo-liberal climate, laid on households. Take for example waste: municipalities formerly came to pick up your waste. Now households are expected to deliver increasing amounts of waste themselves to dumps. But one can, according to Jarvis, also think of home care, volunteer aid, or too little capacity in day nurseries.

Jarvis also sees the restrictions created by opening hours regimes. Much coordination is needed within households. Part of the time scarcity is a result of government decisions, of the moral climate, and of status elements; you attain a higher status by explaining that you are very busy.[10] She asks for a good theory on daily coordination and she concludes; 'do we care sufficiently about the consequence of escalating inequality, congestion, pollution and uneven development, to invest in public solutions to private coordination problems, when these threaten social cohesion and environmental sustainability?' (Jarvis, 2004, 14).

How many trips now made by car to arrange the different activities in set timeframes do have alternatives? In two related articles the question whether trips by car can also be made by other transport modes is studied. In 2002 from Dijst and Kwan wrote *Accessibility and Quality of Life: Time – Geographic Perspectives* and Dijst, de Jong and Ritsema van Eck; (*Opportunities for Transport Mode Change: An Exploration of a Disintegrated Approach*). The conclusion of the studies, carried out in the Rokkeveen neighbourhood, in Zoetermeer, in the Netherlands are interesting. Nearly half of households could make the same chain of trips in the same time by other transport modes. Sometimes the order of activities has to be changed, which is not always possible. Older people have the most possibilities of change, but also two-earner families come near the 50 per cent mentioned.

This conclusion also means that 50 per cent of households are not able to change to other transport modes. The study concentrated on the physical possibilities; aspects like convenience and freedom have not been taken into account. Looking at all possibilities, car reliance in Rokkeveen could diminish by 30 per cent. When all trips have to be done by car the car reliance should grow to 60 per cent. To reduce car reliance in particular better routing of public transport, better connections of the time tables of public transport modes and more use of delivery services is necessary.

---

10   On this theme Gershuny (2005); *Busyness as a Badge of Honour for the New Subordinate Working Class.*

In the region of Avignon in France, Genre-Grandpierre and Josselin studied, in *Exploring the 'Tensions' in the Urban Mobility System* (2008), the vulnerability of the car-oriented transport system. This system relies more and more on the speed of transportation, and hence on the private cars. They study the possible tensions that arise with lower speed through extra traffic. Not the central city, but the surrounding suburbs in a ring of 12 miles around Avignon, will then face the greatest problems. They suggest that creating more resilience in this urban mobility system appears necessary.

To end this motive, time pressure can also arise from other aspects of modern life. In *Driving by Choice or by Necessity: The Case of the Soccer Mom and Other Stories* (2005) the authors Handy, Weston, and Mokhtarian signal that car drivers often do not choose the shortest route or the nearest facility. A percentage of journeys taken and of the vehicle miles travelled is in their eyes 'beyond what is necessary'. This relates to; taking a longer route, taking a more difficult route, choosing for a further destination than the nearest facility. In this respect people create time pressure and time scarcity themselves. Partly this is foreseen and accepted; partly this is the result of wrong information or awkward planning.

To conclude: this whole complex of time schedules, need for coordination, many wishes, time scarcity, stress, hurriedness, does not look very sustainable. The wish for sustainable development seems to be far away in this standard modern practice. Relaxation is a direction not sought after. To put this in perspective; many, often older, households, but also the unemployed, have so much leisure time that the aforementioned situations are outside their realities. And many households, especially in the mentality group of the traditional citizens still lead local lives, with most activities in the vicinity of their homes.

Points for attention are: the social exclusion of some groups (see Chapter 6), the possibilities for substituting car journeys with journeys of other transport modes, and, above all, creating a time policy aimed at diminishing hurriedness, coordination problems, and stress in our modern risk societies!

3.4   Protection

The car is a solid and robust object. Drivers and passengers can be protected by cars, and the car can work as a defence against the unknown, and against real dangers of insecurity and criminality. In modern risk societies people have the chance and the capacity to exclude or not to allow many undesirable deeds, expectations, views and other human beings. They can, simply stated, put them outside their gates. The car can be of great help in this.

In Section 3.3 I concluded that many people face time scarcity. People feel hurried, stressed, or vulnerable. In these circumstances, they will not allow themselves to be confronted with situations that can raise anxiety. They are happy to avoid such situations and have the capacity to do so. The car can be quite instrumental in achieving this in at least three situations:

The first is *the exposure to traffic*. The car is the most solid of the individual transport modes for passengers. You feel safer in a car than walking, biking, or motorcycling. The car can protect you from the negative or discomforting aspects of other road users.

The second is *the exposure to the rules and conditions of public transport*. Public transport is more difficult than the car; you mostly need to travel some distance before you reach a station or a bus stop, and the same holds true on the arrival. Public transport is more difficult with luggage, and is at its core social, meaning that you will be confronted with other people, a confrontation that can be nice and rewarding, but often this will not be the case. To use public transport professionally you have to be aware of time tables. For infrequent users public transport is difficult to read. These all are circumstances that complicate life where simplification is sought.

The third situation is *the exposure to life in public spaces*. Formerly public spaces were the living areas for the whole population, for all inhabitants, a terrain and an area for meeting different lifestyles. That was not always easy, everybody, rich and poor used the same spaces, and subtle rules of behaviour came into existence. Public spaces were not empty, and offered danger and chances at the same time. People could not easily withdraw from public spaces.

With growing prosperity and with the arrival of private motorised mobility the presence in public spaces becomes a choice. Most people choose to avoid these ambivalent spaces. There are less real users and, more important, the composition of the users' changes, from everybody to more specific, mostly lower income, groups.

The reaction to each of these three exposures contains a common ground. This common ground can be described as the wish – when the possibilities are available – to choose for one's own, individual, sphere. This is a choice of not connecting, against social bonds, against the collectivity or against togetherness. This looks like a free choice but behind this choice elements of anxiety and mistrust can be noted. Anxiety and mistrust are not easily communicated; most people will argue that the choice of car use over the choice of public transport is a rational one.

There are only a few articles about the choice of avoiding exposure to traffic. There is more literature on the avoidance of public transport and the literature on the fate of the public space in western risk societies is ubiquitous. In 1976 Sennett published *The Fall of Public Man* and set the tone for the debate. His thesis is that in earlier times people had the ability to show specific behaviour for public spaces, with specific codes and rituals. You kept your distance, and created encounters. This specific culture has been lost. The literature on public spaces can be read as a permanent debate with Sennett.

The car is the instrument for creating avoidance, in all three situations. The car is the winner in individualised traffic, is the alternative to public transport, and can cut through public spaces, without connecting. There are some side effects with these avoidances.

In traffic it is clear that *safety for the car drivers can lead to lack of safety for other road users*. Roads can become 'battlegrounds for limited space' (Beckmann, 2004). Cars can qualify in a world of anonymous engines, can dominate space, and this feels safe to their drivers and passengers but it does not feel safe to more vulnerable road users. Something changes when people, having previously walked or cycled, take the car. Beckmann tries to explain this change in *Mobility and Safety* (2004). In his vision mobility is becoming important for understanding modern society. The car is a metaphor for mobility. The car can be seen as inherently ambiguous; the car avoids danger, and creates danger, the car is there for mobility, but leads to immobility, by using space for others (for example streets for children's play).[11]

And the car is part of a hybrid. On the road the driver and the car form that hybrid. For other participants in traffic the driver–car hybrid can be seen as 'a monster in a metal cocoon', a cyborg with human and non-human characteristics. With the introduction of more driver support systems this view of car and driver grows stronger. In cars the changes in the interaction driver – cars are central, from a greater distance the car-driver becomes less human in its steering. This hybrid raises anxiety, and that anxiety stops the moment you yourself form part of a hybrid. Beckmann shows that the former independence of the driver will disappear. Drivers use techniques, and these techniques of driving support are a step on the way to controlled car mobility, by further hybridisation car-driver.

Aggression is a daily worry for road users. A part of this aggression seems to come from the tension between the idea of mastery over traffic on the one hand and the social and rather impractical elements that are part of traffic in densely populated areas.[12] In *Roads for Change: Changing the Car and its Expression* (2004) Redshaw draws attention to another aspect. People feel safe in their cars, and strong. Many metaphors for car sales and many advertisements refer to racing. The expectation is created that racing, with its freedom, its speed, and with the art of navigation is the superior form of driving.[13] To be prevented from this 'superior' form can lead to aggression in traffic, to so-called 'road rage'. Redshaw considers it necessary to communicate proper and wise articulations and associations around the use of the car. Showing that the car is an instrument for freedom creates other behaviour than defining the car as a great helper in giving help and support to friends and family with health problems (see also Balkmar, 2007).

Moeckli and Lee show, in *The Making of Driving Cultures* (2005), the contours of a desired driving culture. In the US there are many traffic-related victims, but there is no debate. And nearly 65 per cent of the victims are the result of aggression-related driving acts; perhaps because of the familiarity of car travel

---

11   For a broad and sometimes philosophical book about car, car use and safety see Packer, *Mobility without Mayhem* (2008).

12   For further reading see Van der Bilt, *Traffic* (2008).

13   Redshaw elaborates this theme further in, *In the Company of Cars: Driving as a Social and Cultural Practice* (2008).

and its instrumental role in our daily lives, crashes are accepted as unavoidable consequences of the convenience of car travel (Moeckli and Lee, 2005, 65). Moeckli and Lee also offer an explanation for 'road rage': 'the pseudo-private space afforded by the 'metal cocoon' of the car permits some drivers to act against social order, with the effect of justifying violent and dangerous driving by denying the humanity of the other roadway occupants'(Moeckli and Lee, 2005, 67).

Mesken signals in *Determinants and Consequences of Driver Emotions* (2006) that aggression and anger in traffic in The Netherlands are primarily related to the behaviour of other road users. Making traffic systems more predictable and minimising the interactions between road users will help to diminish aggression in traffic.

In recent decades people have moved away from public transport in most western societies. *More and more households are no longer connected to public transport*. Bus transport outside the conurbations, develops into a transport model for well-defined groups like students, the school-going masses, members from non-car households and elderly women. Buses are not a service for a broad spectrum of clients anymore. The sphere and tone of this branch of public transport is changing. The chances for regeneration of bus transport as an alternative mode of transport for a broader client spectrum are becoming rather low.

In 2003 the Dutch social research institute ITS published *Verknocht aan de Auto?; een onderzoek naar determinanten van vervoerwijze.* Higher-educated persons, travelling greater distances in rush hour periods are the most public transport minded (read, train-minded), lower-educated people, travelling shorter distances outside rush hours are the least public transport minded. At least in the Netherlands. There are far more people with a car preference, than with a public transport preference. Car lovers are far more negative about public transport than public transport lovers are about cars. Public transport gets the lowest marks from people who never use it. The least inviting aspects of public transport travel are the crowdedness, the time pressure felt, the poor punctuality, and the situation that it is laborious and time-consuming. Recently, social aggression is also seen as a pitfall of public transport. Cars do not have these unattractive elements.

People sometimes dislike buses. In Edinburgh, Stradling did a study about the negative aspects of bus transport under the title *Eight Reasons People Don't Like Buses* (2002). He identified four blocks:

1. The first and most important block is irritation with other people. It covers drunken passengers, the use of mobile phones, vandalism at the stops, anxiety for personal safety in the evenings, bad driving by bus drivers, and more widely, the behaviour of fellow passengers.
2. The second block concerns the laboriousness of the payment system in bus transport.
3. The third block covers anxieties about arriving on time, in the journey, but also at the bus stop.
4. Finally, the luggage is always a difficult element.

From a factor analysis the greatest irritation was 'unwanted encounters'; you just want to make a trip and, unasked, you are confronted with all sorts of persons and situations that you do not want to deal with, and that confront you with the harsher and bleaker side of public life. In your car you are not confronted with this unwanted encounters, you can close yourself off from these kinds of experiences. Parents, and certainly middle-class parents, do not want to confront their children with the realities of life.[14]

In *Better for Everyone? Travel Experiences and Transport Exclusion* (Hine and Mitchell, 2001) the focus is the experiences of the 'public transport captives', people who have to take public transport because they do not own a car. Public transport is difficult with young children, and with luggage. As an older woman says, 'it's carrying the luggage, that's the problem. Public transport is not really geared for luggage' (Hine and Mitchell, 2001, 323). Some of the necessary journeys are not made by the 'captives', and other journeys are adapted to what is feasible in their condition. An elderly couple, walking slowly, went for their shopping three times on the same day by bus to the supermarket. Only in this way could all their purchases be transported. Captives also raised the topic of public transport being unpredictable, especially in the evenings, and there are many complaints about the behaviour of the bus personnel. The article paints a picture of a service with a weak innovative attitude, knowing very well that they transport captives. With a car, again, you do not have to be confronted with these disappointing experiences.

In *Trip Chaining, Childcare and Personal Safety* (1996) Bianco and Lawson summarised US research on the experience of safety in bus traffic. Non-users have greater anxiety than frequent users and users are more concerned about time before the bus ride than during the bus ride itself. Women are more anxious than men. A number of women do not travel by night by public transport.

They feel vulnerable in public spaces. *The public space becomes an area to travel through, and a space for traffic, at least for a majority of households.* Many households consider connections in public spaces – social at a distance, but physically near – a combination to avoid. Other households, mostly from the lower income groups, see the same public spaces as their living spaces. A spatial 'sorting out' of different social groups arises in public spaces. Higher and middle-class households have friends in a broad and dispersed network. For them, the vicinity is not very important. Solidarity and connectedness is for these households primarily related to places where friends and relatives live, and less to their own neighbourhoods. They live in those neighbourhoods but travel quickly in and out by car. Solidarity with neighbours seems to erode.

In 1999 the Dutch Social Research Institute (SCP) published *De Stad op Straat*. In our society, the authors conclude, the 'placeness' of societal functions has diminished. Classical and locally bound citizenship is diminishing, and the influence of governments and authorities is growing. Subcultures are flourishing in public space. This all has an effect on the characteristics of public spaces. The elite do not

---

14   Whether this form of escapism is useful is not a theme in this book.

live in the cities, at least not in the majority. This elite is oriented towards the world and not towards their own cities. The public spaces in cities are no longer within their 'responsibility zone'. Finally, public spaces become emptier and become inhabited in the majority by economically weaker groups, and youth subcultures.

For Flanders, Laermans paints a clear picture in *Stedelijkheid in de veralgemeende moderniteit* (2006). He sees a de-urbanisation of modernity. Cities are no longer the bearer of that modernity. Mass media have become the transmission post of new modes and cultures. Public spaces heterogenise; different lifestyles can all have their pockets, small areas, islands. On the other hand, public spaces homogenise; everywhere the real bearers of the modern public spaces are the weaker social groups (see Boomkens, 2006). Design, layout and the spectrum of shopping opportunities are becoming standardised in our European cities. The process of modernisation has winners and losers living in different cultures. Lower educated autochthones face the same problems as the allochthones, but the first group uses the second as scapegoats. People with fewer choices remain in an empty, and more aggressive, public space. Middle-class citizens avoid these public spaces, and drive through them by car.

In *Literature Review of Public Space and Local Environment* (2001) Williams and Green concluded for the UK that public spaces, parks and greener areas are diminishing in quality. The five most important reasons for this are: traffic, small polluting firms near public spaces, anti-social behaviour and criminality, bad design of public spaces and privatisation of public spaces. Too much traffic through public space leads to avoidance, to loss of social control and in the end to small crime.

Pernack (2005) paints a picture for Germany in *Offentlicher Raum und Verkehr; eine Sozialtheoretische Annaherung.* Public spaces are in crisis; people do not know how to behave there. The 'self' and anti-social behaviour have become the norm in society. Human beings like to feel intimacy, but public spaces cannot offer intimacy. People no longer know how to behave in dialogues with fellow human beings with other orientations and lifestyles in an area that belongs to nobody. The public space is nowadays not a 'rules and codes governed' area; '*der Offentlichen Raum verliert seine Identitatstiftende Funktion. Er wird mittels Verkehrsmittel oder ITC uberbruckt und fragmentiert*' (Pernack, 2005, 25; 'The public space loses its identity forming function. That space is fragmentised and bridged by car or ITC'). Here the role of the car comes into view. The car made it possible for people to avoid public spaces; '*Das Automobil ist das ideale Mittel um den zu einer Transitzone verkommenen Offentichen Raum zu uberbrucken*'(Pernack, 2005, 30, 'The car is the ideal transport mode for crossing the transit zone that formerly was known as public space'). While part of the population, the middle and higher classes, retreat from public spaces, the lower classes intensify their presence. Pernack ends;

> *immer mehr Fluchtwege mentaler wie auch physischer art sind entstanden, um sich den urbane Erfordernisse zu entziehen. Fur viele Menschen stellt es auch eine immense psychische Belastung dar, permanent auf sich verandernde*

Situationenen zu reagieren und sie aus eigenen Ressourcen heraus zu bewaltigen. Viele Menschen sind dazu nicht fahig oder gewillt und ziehen sich zuruck. Unlesbare Raume warden gemieten, und sichere, lesbare gesucht (Pernack, 2005, 40).

many escape routes have been created to withdraw from difficulties in public space, mental routes as well as physical routes. May people find it burdensome to react to ever changing situations with their own straightforward interventions. Many people are not so skilful in public and withdraw. Spaces and areas that are difficult to read are avoided, easily readable areas are sought.

In conclusion, the picture remains mixed. The car gives protection to its users and has played a role in the worsening of elements of the living conditions of already vulnerable traffic users, captives of public transport, and more general to the outlooks of the urban lower classes. Cars can play a role in the further division of society. This was analysed for the transport story of Hurricane Katrina in New Orleans (Sanchez and Brenman, 2007; Creswell, 2008). The carless people really were trapped!

3.5   Freedom

Freedom is probably the oldest motive for car driving. The first motorists recounted enthusiastic stories about their experiences of speed, which immediately gave a feeling of freedom. Speed, power, beauty, adventure: these themes are central in books[15] about the history of car driving. The demand for speed, the wish to explore new frontiers, are deeply entrenched in modern mankind. But the earth has been discovered, there are no undiscovered places left. Some areas are empty and dangerous, but not as dangerous as they once were. Romanticism, the urge for adventurous spaces, is rather difficult to satisfy. The car has become a normal consumer item, and is no longer a help in *exploring new frontiers.* Part of this role has remained in leisure.

In *Three Ages of the Automobile* (2004) Gartmann constructs the development of the American car from a historical perspective. He identifies:

- the age of class difference; until 1925. The elite could afford a special car, for the middle class a functional and simple car came into view. This was accepted.
- the age of mass individuality; after 1925 people want more variety. They want compensation for their boring work, and consumption can bring this compensation. Cars are made for them with the same frames, but with different end models. The production line can remain more or less the same.

---

15   See, for example, Staal (2003), Filarski and Mom (2008), and Van der Vinne (2010).

- the age of diversity in subcultures; at the beginning of the sixties prosperity has grown so much that a real demand for really different models arises. Car manufacturers are going to deliver these models, but the deliveries need new organisations of the car production. Import of cars from other countries becomes important.

The driving force behind this development is the wish of car buyers to present their own individuality via the car. Gartmann notices that the laws of the market mean that people have to realise their deep-seated need for identity and autonomy via consumption. An ultimate form of this 'compensating consumption' is the car.

In more recent years the car *gets a role to create freedom for people* – the young, for the first time, or the elderly, still – *to be able to participate in modern society*. Youngsters, especially, feel that without a car they can only participate in part in the opportunities of modern society. Getting your driver's licence is a modern 'rite of passage'. For older people the car allows them to still take part in society. They do not have to depend on other people (family, friends, taxi drivers) or on public transport. Giving up driving is, for men, a very important step.[16]

Car plus gadgets *create possibilities for self-expression.*[17] This seems important for people who do not have many other opportunities for self-expression. This third element of the freedom motive is the fact that the car can be seen as something that is unique, and really yours. The car has, together with a number of hobbies, the capacity to offer people, who are not able the find their individuality in other spheres of life, a form of compensatory individuality. They can, through the purchase of a unique car, through developing the skills to repair or build cars, through focussing on their car as an art object.[18] More broadly, cars can be a source for national pride and for national identity (Edensor, 2004).

The final element of freedom is the freedom the car can create in *the time in-between*. At your start location there is pressure, at your destination also. In-between, in your car, you are on your own. The car journey allows you to be by yourself, for a shorter or longer time. Many people love this 'in-between' time. It can be filled by looking at the landscape, by talking, by listening to your favourite music, or your favourite radio programme. There is a need to be on your own, to have some peace and quietness in a busy life in a busy society. The car now has so many gadgets, that you can more-or-less feel at home. Car driving for many car drivers contains elements of beauty (see Borden, 2005), or mastery. Many people

---

16   I will discuss this further in Section 6.3.

17   In *Cars and Behaviour: Psychological Barriers to Car Restraint and Sustainable Urban Transport* by Diekstra and Kroon (2003) the authors stress this argument. They see the car as an element to impress, to show your macho-maleness, as a toy for old boys. In their view, a man undergoes a personality change when driving a car!

18   Examples are to be found in the Car Care Project by the British sociologist Dant (2004b).

like to have a buffer between their work and their home, they like to drive a little bit longer to feel the relaxation. Your car feels more like home already than like work.

In *Automotive Emotions; Feeling the Car* (2004) Sheller presents an overview of the emotions the car evokes. Car driving gives an impression of liberation, support and social connectedness. The car is 'part of a pattern of sociability'. The car seat is integrated in daily and weekly routines. Sheller warns that 'anti-car ethics' relates not to the need to keep the social mobile for family and friends networks. Many people consider the car too comfortable, too pleasant and too exciting to abandon car use, although they know all the environmental, energy, public space and traffic related problems. Her main question is; 'For those who have become so deeply attached to their cars and to the physical, cultural and emotional geographies that have become natural within car cultures, how easy will it be to give up this part of the self, the family, friendship and kin networks' (Sheller, 2004, 225).

In the literature on motives for frequent car use sociologists, transport professionals and human geographers dominate. Philosophers sometimes have given thought to cars and car use. The core theme of the French philosopher Virilio is speed.[19] The also French philosopher Augé (1995) notices the capacity of the car to create non-places and transit landscapes. Lomansky, an American, is a third philosopher on car use. In *Autonomy and Automobility* (1995) he states that car mobility is related to autonomy, to the capacity to define your direction yourself. The car facilitates this autonomy and stands alongside book printing and the computer in this respect. The car makes us less dependent on pre-organised arrangements; we can choose for ourselves; 'Detroit (city of the American car industry) has done more for the liberation and dignity of labour than all Socialist Internationals combined!'(Lomansky, 1995, 17). In his opinion, in travelling by car your experience level grows; the car is 'the quintessential range extender'.

In a number of studies the car is described as more than just a tool. In *Travel for the Fun of It* (1999) Mokhtarian and Salomon describe the pleasures of driving by car. They have their doubts of transport being predominantly 'derived demand'. There is certainly driving just for the driving. They look at excess travel', which is about driving unnecessary miles on routine journeys. People want something more adventurous, just a little detour, want to get away from daily rhythms, and take with full consciousness a longer route. Many people value their car time as a buffer between work time and house time. On average, Americans would like to travel for 17 minutes on their drive from work to house, and vice versa. Many people like longer journeys (see also, Ory and Mokhtarian, 2005).

In *Pour faire face a l'automobilité: des transport en commun plus ludiques?* (2008), Meissonnier looks at the reasons behind the poor results of a mobility management initiative taken by a French employer. Very few people joined the scheme, everybody kept driving although there was congestion on the roads near

---

19   For an introduction to the work of Virilio, see Armitage (1999). For Virilio; *The Open Sky* (1997).

the industrial zone. He gives a number of reasons: most employees hated to be on a schedule for their journey, felt already scheduled in their working life. But more important is showing that you have a nicer car than your director. That car gives you also a feeling of control over your own life. And the car creates flexibility; 'La voiture, c'est la surf de la ville' (Meisonnier, 2008, 44). He concludes that these reasons make it necessary for forms of joint or public transport to create more fun, present their services in a way that creates more pretty or happy feelings.

Laurier works with images in *Habitable Cars* (2005). He shows the possibilities of a phenomenology of car use,[20] as he describes a journey to school, also taking other parents' children. For outsiders this looks like a simple trip, he shows how much advanced thinking, planning and hidden precision has been involved. But a rather dry description of this trip does not show you 'the many mutual obligations, the flavours, the work, the trust, the aid and the generosity' (Laurier, 2005, 5). People do more with cars than just driving. Each trip is an enterprise on its own. And the car becomes in a journey as described, a sort of living room. This all means that your car feels like home.

There is also freedom in defining your time. Lyons signals in *Future Mobility – It's about Time* (2003a) that travel time is fast becoming activity time. The unproductiveness of a car journey is ending. Work comes into the car, by mobile phone, by laptop, and especially in congested situations. The car then becomes a mobile office, thus creating difficult questions. Is the working time in the car counted as work? There is an intermediate zone, which is not the case for train passenger. They can work on their journey. Car drivers have to drive, and from a safety perspective it is probably better to see that as their only activity. Even when working, driving cannot be far away. Also this intermediate situation gives car drivers a feeling of freedom.

The car can also give freedom to different population groups. In *Consuming the Car* (2002) Carrabine and Longhurst present the function of the car for youngsters between 15 and 18 years old in two neighbourhoods in Manchester. They identify a significance orientation and a problem orientation. In this last orientation elements like road rage, joyriding and vulnerability to car advertisements come into view. This is the showing-off side, the thrilling, the excitement area. The significance orientation is about the role of the car in being able to join greater society with all its chances and possibilities. Youngsters negotiate with their parents and peers about organising car traffic. They are anxious about becoming outsiders among their friends, when they cannot join parties, activities or events. The majority want a car as fast as possible, to be able to participate and also because the car is 'a protective shield in the management of risks going to the city' (Carrabine and Longhurst, 2002, 190).

20　There are only a few phenomenological oriented researchers in mobility studies. See also Laurier's *Doing Office on the Motorway* (2004), or; *Living on the Motorway* (Delalex 2003) and *Taking the SUV to a Place it has Never Been Before; SUV Ads and the Consumption of Nature* (Aronczyk 2005).

Older people face other problems. In *Automobility among the Elderly* (2002) Rosenbloom and Stahl clarify that each new age group of elderly will be less public transport-oriented. Older people up to the age of 77 drive a lot. For older people the loss of their capacity to drive is a great problem. Many older people, especially men, persevere in driving, sometimes crossing the line where driving becomes dangerous both for themselves and for other road users. In the US, in traffic models insufficient attention has been paid to driving in old age. The car dependence of many older people is relatively high; they postpone the decision to stop, but adapt their driving behaviour, and travel more in off-peak hours.

In a French study from 2002, *La pauvreté peri-urbaine: dependence locale ou dependence automobile*, Coutard, Dupuy and Fol compare the use of the car in two rather poor neighbourhoods near cities, in France and England. In the French situation a third of the households have no driving licence and no car. It mostly considers big, extended families, living near each other and supporting each other. Shops, partly subsidised, are nearby, and the same holds true for a spectrum of government services (including reasonable public transport). Only a few people leave their neighbourhood or its immediate surroundings. There is a low mobility need. Everything can be found in the neighbourhood, but they are more or less stuck there. This is 'dependence locale'.

In England another picture can be seen. First, non-car use is far lower, 22 per cent. There are no shops and no government services in the neighbourhood. Without a car you are really stuck. You then have to use the infrequent public transport and expensive taxis. The car, rather expensive in the small budgets of lower income families, gives exactly the ticket they need to participate in society and to be able to buy groceries. A car is in these circumstances a necessity. This is called 'dependence automobile'. This study shows how differences in political style, attitude, and decisions lead to generating or avoiding car dependence.

3.6   Habit

This motive looks so obvious that is often forgotten. Over time, people develop standard patterns for activities. They do not think about every action they make. These standards patterns are responsible for the situation that new alternatives for an activity are not seen or not picked up. Sometimes alternatives are assessed but pretty easily rejected.

In general it is useful that routines, following the accustomed standard, arise. It would cost too much energy to do otherwise. Many things in life are done more or less automatically. This holds also true for the choice of transport modes. When a transport mode delivered the services in previous situations, there seems no need to change. 'The same again' is then a normal and understandable practice.

For many actions a routine has developed. This means that this action is always done in the same way, but it can also mean that a few alternatives will be considered. Routines are responsible for the situation where people show at first little openness towards chances of improvement, or new experiences. There is some inherent conservatism here. Habit as an explanation is discussed in *Residual Effects of Past on Later Behaviour* (Ajzen, 2002). The author argues that the frequency of a line of conduct is a good predictor for future behaviour. However, customary behaviour is not fully automatic, but rather semi-automatic. Usually you will shake hands, but you stop and think when somebody has wet hands. In Ajzen's eyes habit cannot be placed on the same line as the frequency of former behaviour. What can be said is that behaviour is rather stable in time: 'Strictly speaking, the association between past behaviour frequency and frequency of later behaviour, by itself, merely demonstrates that the behaviour in question is stable over time' (Ajzen, 2002, 110). The reasons for not realising new behaviour are in Ajzen's analysis not habits, but unrealistic expectations, weak or instable attitudes, and inadequate planning for the new, desired, behaviour.

In most households the car is the first transport mode to be considered. The car can be used immediately. When a destination was reached by car in earlier days, why not take the car again?

Also an important element is that ownership leads to use. Once a car has been bought a household changes situation; very often once there is a car, the car is used for almost all journeys. Selectivity in use mostly comes later. Aarts and Dijksterhuis report in *Habits as Knowledge Structures; Automaticity in Goal-Directed Behaviour* (2000) their research to automaticity in choices of transport modes. Their idea was that 'when the destination and the purpose were known (e.g. going to the city to shop) that automatically (without conscious thinking) the transport mode would be activated that brings you to that destination' (Aarts and Dijksterhuis, 2000, 55). They noticed more automatic behaviour than Ajzen. But they also recognise the role of planning in breaking routines.

An equilibrium between routine and openness for new alternatives would be best practice, but this is not easy reached. The car supports routines. Harms reaches a pragmatic definition of routine in *From Routine Choice to Rational Decision Making Between Mobility Alternatives* (2003). Routine is an automated strategy to cope with the environment in order to reach a desired purpose or destination. It deals with stable contexts and satisfactory achievements. Strong routines make people blind for new options. She acknowledges three types of blindness; cognitive, motivational and uncertainty. There are moments when breakthroughs through these 'blindnesses' can be reached. Once a routine has been broken, and another decision has been made, more is possible. She illustrates this with an example on car-sharing in Switzerland. As routines are weakest during changes in circumstances, she advises looking carefully at these moments for change.

To use public transport you need to know how it works, you need to know the timetables. The car is just there. Selective use will possibly develop when, for each trip, you can immediately see the price of that trip. This is nowhere near the case. The situation is the other way round. Because of the fixed cost orientation in car pricing the costs of a car trip are perceived as far lower than the price of a public transport ticket.

Does this mean that routines cannot be broken? Certainly not! There are tipping points. Behaviour is relatively stable, but there are moments when breakthroughs are possible. Gladwell developed a theory for correcting routines, or changing actual routines by freer choices or a new routine, certainly of use in *The Tipping Point* (2000). Klöckner searches for these important moments in life in *Das Zusammenspiel von Gewohnheiten und Normen in der Verkehrsmittelwahl* (2005). He mentions starting a new education, relocations, getting your driver's license, going to a new school, buying a car and starting a first job, or a new job. Especially at these tipping pints there are 'windows of opportunity' for a change in transport modes.

To end, I will now focus on three practical studies about habits and mobility. In *Choice of Travel Mode in the Theory of Planned Behaviour; The Roles of Past Behaviour, Habit and Reasoned Action* (2003) Bamberg et al. follow Ajzen in his rejection of habit as an explanation. Into a relatively stable transport situation they introduced a new element; a free ticket for public transport. There was a swift change in this situation. They concluded that the so called automatic behaviour was not very resistant, not very resilient, in a situation where a real alternative was offered.

Fuji and Kitamura present in *What does a month's free bus ticket do to habitual drivers?* (2003) an experiment in Tokyo. In a situation where a highway was out of use they gave habitual car drivers a month's free public transport. Car drivers made 20 per cent more use of the buses. At the end of the month, and after reopening of the highway, the decrease of car use was significant, but the growth in bus transport was not significant.[21]

Burbridge et al. conclude in *Travel Behaviour Comparisons of Active Living and Inactive Living* (2005) that within car reliance in the US huge differences can be seen. Also, differences in habits between active and inactive car drivers can be seen. Active drivers will, next to their car driving, also choose biking and walking, inactive drivers remain with their cars. This last group drives 16 minutes more each day.

Active car drivers do have thresholds for the active transport modes: For walking the threshold is 2.3 kilometres, for cycling 5.8 kilometres. Above 9.7 kilometres the car has taken over almost all passenger transport. Burbridge concludes that low densities diminish the chances for substituting car driving with other transport modes.

Concluding, the experts disagree whether habit can be an explanation for behaviour, but it is at least clear that in stable situations past behaviour is good predictor. On an individual level a semi-automatic pattern of behaviour seems nearer to reality than a fully automatic behaviour.

Routines are weakest at tipping points. At these well-defined moments change in routines can be expected or organised. Experiments show that changes in habits are possible, although it has to be accepted that a number of these changes will be temporary.

––––––––––––––––––––

21   A somewhat comparable experiment was done by Thorgensen and Moller (2007). Here, after ending the experiment, car drivers went back to their cars.

3.7   Individual motives for car use: Some first conclusions

I identified that probably half of car drivers have a real pro-car attitude. For the other half the car is seen as a useful commodity. Having a real pro-car attitude correlates probably with:

- emphasising convenience and comfort
- avoiding contact with unknown people
- cherishing the feeling of freedom the car can give
- seeing car driving as a habit.

A very important motive for frequent car use is the need to be flexible as identified in Chapter 2. Another very important motive is protection from the dangers to be found in traffic, on public transport and in public spaces. This motive was also identified in Chapter 3. It looks as though:

- convenience and comfort
- avoiding contact with other people
- cherishing the feeling of freedom
- cherishing the status elements around the car
- creating possibilities to participate in society
- habit

are the real individual motives for car use.

To conclude the analyses of Chapters 2 and 3, I will end with a scheme of driving forces for car use, identified so far.

Table 3.1   Scheme of driving forces for car use, identified so far

Individual attitudes and motives	Characteristics risk society
Convenience and comfort	The urge for flexibility, combining tasks ain tight timeframes
Avoidance of contact with strangers	Geographical spreading out of activities
Cherishing the feeling of freedom with a car	Fears, feeling vulnerable, seeking protection
Cherishing the idea of status related to the car	Implicit expectations in the social sphere
Habit in choice of transport modes	Everything on appointment
Creating possibilities for participating in life's experiences	

Chapter 4

From Frequent Car Use to Car Dependence

4.1 Big societal stories, with Doing It Yourself as an example

A central concept in this book is the big societal story. Big societal stories clarify during a particular time the dominating lines of thought in a society. These stories also legitimate the choices on offer. In modern big societal stories the characteristics of modern risk society (Chapter 2) and the dominating motives of the individual members of this society (Chapter 3) come together. Big societal stories look, for a longer time period, relatively stable.

Big societal stories are structural stories. For mobility studies the concept of structural stories is coined by Freudendal-Pedersen (2005, 2008). Structural stories are short, and can be seen as standing truths, as obvious clarifications of activity patterns. They explain the present ways of acting, and the situations that led to these ways of acting. In *Mobility and Daily Life: between Freedom and Unfreedom* (2008) she works with three structural stories; 'when you have children you need a car', 'the train is always too late', and 'the car offers you possibilities that no other transport mode can offer'.

Car use fits in big societal stories and car use makes new and other societal stories possible. Big societal stories are not static, they come and go, but as stated, for a time they look rather stable. Changing big societal stories takes time but stories die. Two examples can be given on mobility.

The first is hitchhiking; very popular in the 1960s and 1970s, now a marginal activity. The story has been lost due to the growth in the number of cars, to the fact that far more youngsters drive at an early age, to the cheaper possibilities of public transport, especially for students, and to the changing perception of danger and risks in picking up strangers.

The second is the story of the moped. Mopeds were once very popular as a mode of transport: there was even a moped culture. Mopeds were an important means of transport for poorer households, and later for youngsters. Now few mopeds remain. Part of their role has been taken over by the scooter, but the bigger part has been lost. Reasons for the loss of this story are probably: poor press regarding the safety of mopeds, the possibility, with growing prosperity, for poorer people to buy cars, and successful advertising campaigns for modern bikes.

Big societal stories are built by the driving forces, but these stories contain more: emotions, fears, expectations, and dilemmas are constituting elements in these big societal stories. I will describe a newer big societal story with a strong relationship with frequent car use.

The rise of Do It Yourself (DIY) in our modern societies is very interesting. DIY was, for a long time, a marginal activity. Now over 60 per cent of households in Western Europe engage in DIY, and in these activities in the Netherlands more than 4 billion euros are spent annually. With DIY two types of motive are important: global and specific.

First, the global motives: Growing prosperity has led to higher budgets in most households and with the arrival of work-free Saturdays in the 1960s, more available time. More recently, household dynamics became far more important.[1] People divorce more easily, get new partners, keep two houses, or change to one house, but with new wishes and needs. Building, rebuilding, changing the divisions of labour within the household, all have become more important. Moving to another house when changing work location happens less, due to the structure of housing markets, but also due to the fact that in dual-income households the location decision is a compromise between two travel partners.

In recent decades the ratio between the budgets of households and the wages of building professionals has changed. Today professional wages are relatively high. The last global motive is that most people now do sedentary work – there is a high demand to spend free time actively.

All these motives can be found in the literature (see Brodersen, 2003; Moisio, 2007; Watson and Shove, 2005 and 2006). These global motives are a first explanation for the growth of DIY. Growing prosperity, more free time, greater household dynamics, and higher prices to be paid for building professionals make the growth in DIY possible.

That this possibility became a reality finds its reason in more specific motives on the individual level. The first is saving money. Households try to do many alterations on and in their houses themselves, and they employ professionals only for the more difficult jobs. Many people are also sceptical about the quality of those professionals: stories about bad work are ubiquitous. Also, many people have fixed ideas about what the alteration should look like, and feel they can do it better themselves!

The last specific motive is especially relevant for men. In doing DIY you enhance feelings of masculinity. In the US in particular attention is paid to this motive. Foster (2004, 442) states; 'Maleness pervades many aspects of DIY retailing; male customers perceived male customer facing staff to have better knowledge of technical DIY than female employees, though that actually was not always the case.' In Nebraska a thesis was published on this subject. Moisio clarifies, in *Men in No-Man's Land; Providing Manhood through Compensatory Consumption* (2007), two motives for DIY as compensation for a sedentary work life. The first is gender disorder – through DIY men try to reach achievements that are lost in their low status work. This motive is found particularly in the lower classes and the unemployed. The second motive is emasculation, or rather,

---

1 Baker and Kaul (2000) show for the US that extension of households leads to DIY activities.

showing that, although you have a sedentary job, you are still a 'real' man, capable of doing practical things yourself. This motive is found in the higher social strata.

This last motive shows that DIY is a complex phenomenon. It is not only done to save money. There are deeper reasons. Watson and Shove show this in *Doing it Yourself/Products, Competence and Meaning in the Practices of DIY* (2005). They try to clarify DIY; is it consumption, production, or both? And they show that DIY is about more than saving money 'what is missing … is the doing of DIY, the works, the sweat, and frustration of mixing up bodies and their limitations with a diverse array of tools with which to transform a collection of materials to form the effect of a material change to the home, the product of labour' (Watson and Shove, 2005, 4).

Building professionals only come in to do the more difficult jobs. Easy odd jobs are painting, wallpapering, small electrical repairs, and tiling. More difficult odd jobs are stucco work, bricklaying, work on the boiler, the roof and demolition. Brodersen (2003) shows that in North Western Europe households have to work three to five hours to be able to pay for one hour's work by a building professional. At this ratio people only want to pay for the more difficult odd jobs. However, a change appears to be under way. With cheap labour coming to work in Western Europe from the new EU members a fourth layer in the market has developed. You can hire a local professional, a professional from Eastern Europe, somebody from the informal economy, or you can do it yourself (see Buhn et al., 2007; Adriaenssens and Hendrickx, 2008).

We now come to the relation with car use. DIYers spend on average far more time on their odd jobs than the professionals. For most DIYers the work is trial and error and many trips to the DIY stores have to be made. At the weekend the stores are crowded and the products are heavy, with impractical forms and shapes, such as paint, wood, or bigger equipment.

The overwhelming majority of trips to building markets are made by car. In the Netherlands the *Ruimtelijke Planbureau* in *Winkelen in Megaland* (RPB, 2005, Table 1) divided shopping into three motives; Fun, Run and Goal. DIY belongs to the Goal motive and transport for Goal motives is almost exclusively done by car.

Where are the building markets situated? Originally in the city centres, but when their assortment grew exponentially in the 1990s they relocated to the urban fringes, often near highways. There is very little public transport to be found on these locations but they are within easy reach by car, thus being in line with the needs of DIY customers. To quote a DIY entrepreneur; 'The order of wishes of our clients remains stable; at first the reachability, than the parking situation, than the assortment, than the service offered, and finally the price' (Nationale Franchisegids, 2007, 63).

With these elements the big societal story can be written in three structural sentences:

1. Professionals are too expensive, you have to do it yourself.
2. You can do a lot yourself, but there is a line you had better not cross.
3. By doing it yourself you can show you are a real man.

But the real story is greater and story consists of:

- permanent alterations to the home
- realising yourself
- in a very car-oriented manner
- buying elements in peripheral located DIY stores
- and where professionals only do the difficult work.

This car-related big societal story is the result of many decisions. We could have decided to use part of our growing prosperity to pay for professionals. If car densities had not increased so much, building markets would not have been located on the urban fringes. Governments co-created this story, through high growth of wages in the building professions (due to labour safety regulations, and national insurance contributions), and approval of the locations of DIY shops.

This example shows the possibilities of a source policy for mobility. If we want to have fewer car kilometres over short distances (creating higher emissions on CO_2) we have to go back to the original motives for DIY and see whether elements in these motives can be changed. Source policy comes from the environmental policies. In these policies emissions are traced back to the original sources. Attention in these policies went from factories back to global trade patterns. In mobility this form of policy seems non-existent.

4.2   The big societal stories on frequent car use

There are a few big societal stories that together are important in the debate on the reasons for frequent car use. These stories will be presented here, using the material of the Chapters 2 and 3. As stated they are built up of characteristics of modern risk societies, of individual motives, of the driving forces and of emotions and dilemmas. I will present five of these stories.

A story about manners

People in our societies are aware that their modern lifestyles create risks that are, for a greater part, 'man-made'. People know they are vulnerable, but they do not want to face that reality every day, so they adjust and create mitigating arrangements. An important arrangement is the so-called 'community light'; keeping some reservations on fellow human beings, being friendly and polite, but keeping a certain distance. With this attitude it is difficult to ask neighbours, and sometimes even friends, for real help. You need to help yourself, and with this abstract 'doing it yourself' the car is a big helper (for the core of this story see Chapter 2).

A story about time

In modern societies people are expected to be flexible: Flexible on new working conditions, on new ideas. To be able to satisfy all expectations and wishes, a good income has to be brought into the household and both partners have to work. Daily activities and care tasks have to be done, mostly in tight time frames, within explicit opening and closing hours. People have to combine tasks, and they can only meet all the expectations by using the most flexible transport mode, the car (see Chapter 3).

A story about education

In the past children could just exist. Now, particularly in middle class families, they have to become. Parents see raising children as an important task, and take this task seriously. There are many expectations around children and, because parents are anxious about the public space, children are escorted to many locations. Their anxiety is related to 'stranger, danger' and to traffic safety. The concentration of parents on risk avoidance and on interesting hobbies instead of playing in the neighbourhood makes children less streetwise than in the past, and demands planning and organising children's lives. The car is a great help (see Chapter 2).

A story about identity

Living in risk societies is based on abstract rules, functional responsibilities and rituals. We have to fit into these rules. We want to be more than just our working life. We need to feel our strength, our emotions, our capabilities. We want to re-create, doing it ourselves, in our free time. For many people the car is an instrument, in the sense that cars can be instrumental in reaching the areas for authenticity, and in the sense that cars can create identities for their drivers (see Chapter 3).

A story about space

The different activities of modern life do not take place in close proximity; each activity has its own location. These locations have to be connected in order to live a full life and the car is the great connector. Highways become, more and more, focal points for activities. Along the highway you can now work, live, go to the cinema. People grow up with highway experiences. Highways move from the periphery of our attention to the core (see Chapter 2).

All five societal stories need the car. To show this:

Table 4.1   Big societal stories and the role of the car

Big societal story	Role of the car
Manners	Flexibility; risk protection
Time	Realising difficult time schedules, flexibility
Education	Escorting, risk protection
Identity	Easy with luggage, reaching difficult locations, no influence of weather conditions
Space	Connecting different locations and different worlds

One can conclude that the car has become 'embedded' in our modern risk societies. Changing the role of the car means changing the current big societal stories.

4.3   Driving forces for frequent car use

Here the driving forces for frequent car use are analysed and discussed. Driving forces are important elements for the current big societal stories and Chapter 3 concluded with an initial list of driving forces. Some driving forces will be reformulated and they will be split into more societal-oriented driving forces, and driving forces with a strong relation to individual motives. The ten selected driving forces will be presented. Some overlap between the big societal stories and some of the driving forces is identified, and is understandable from the relationship between big societal stories and driving forces.

Table 4.2   The list of driving forces, from Chapter 3, with (in *italics*) the driving forces that will be reformulated

Individual attitudes and motives	Characteristics risk society
Convenience and comfort	The urge for flexibility, combining tasks in tight timeframes
Avoidance of contact with strangers	*Geographical spreading out of activities*
Cherishing the feeling of freedom with a car	Fears, feeling vulnerable, seeking protection
Cherishing the idea of identity related to the car	*Implicit expectations in the social sphere*
Habit in choice of transport modes	*Everything on appointment*
Creating possibilities for participating in life's experiences	

As stated a few driving forces from Chapter 3 will be reformulated.

First 'avoidance of contact with strangers', 'implicit expectations in the social sphere' and 'everything on appointment'. These three driving forces have a common denominator; the wish for 'community light'.

Secondly; 'cherishing the feeling of freedom with a car' will be reformulated to 'the lack of an ethical boundary on mobility; everything that is possible should be done'. This reformulation is chosen because it is stronger, less open and more directed.

Finally 'geographical spreading out of activities'; this driving force has two developments; the ongoing process of relocation of activities, and the deliberate building of highway locations. As this last development is very influential for frequent car use, it looks useful to formulate this driving force separately, and not only as one of the sub developments of the ongoing process of spreading out of activities.

After reformulation the driving forces look as follows:

Table 4.3   Driving forces for frequent car use

Driving forces at societal level	Driving forces at individual level
Geographical spreading out of activities	Convenience and comfort
The creation of highway locations	The wish for 'community light'
The urge for flexibility, combining tasks in tight timeframes	The lack of an ethical boundary on mobility
Creating possibilities for participating in life's experiences	Cherishing the idea of identity related to the car
	Fears, feeling vulnerable, seeking protection
	Habit in choice of transport modes

In describing the 10 driving forces we will start with the forces on a societal level.

Geographical spreading out of activities

Activities, once localised in their vicinities, have in the last four decades moved to other locations. A greater distance has to be travelled to combine activities. The once compact network of locations for related activities has spread out over a wider area. The car can best bridge the distance, and becomes the great connector between activities. The car made spreading out of activities possible and is the best qualified to connect (activities) at different locations.

The creation of highway locations

Many companies grow, and are in need of space. In many municipalities this space can be found by restructuring existing industrial areas. This, however, is a difficult activity, due to the situation that many owners of industrial buildings, warehouses and industrial land, change rather frequently and live and work at great distances from the municipalities.

Municipalities and companies based in them often choose to create new space for industrial activities, for offices and for storage activities. Free space is often found in the vicinity of, or along highways. Industrial activities, tertiary activities but also housing and leisure activities move towards the highways. People working and/or living along highway make use of these highways. They are car-oriented, not in the least because there is relatively little public transport along most highways. People living in the vicinity of highways tend to be oriented to all interesting locations along 'their highway' and are not that much focussed on the city region that they formally belong to.

The urge for flexibility

In modern societies people want to have active lives; they want to work, have to care for their children, are helping friends and family with health problems, and have active leisure activities. Alongside all these activities they have to do daily activities and normal housekeeping. All these activities have to be undertaken in a chain, and in rather tight time frames. Most activities have well-defined start and end times, and professional and specialist services can only be obtained during daytime. The start and end times, the opening hours of shops and services have not changed much in recent decades, meaning that most activities have to be pressed into the hours between eight in the morning and six in the evening. Coordinating activity schedules is essential, as is managing deadlines. Households have to rely on flexibility and speed. Cars and bikes can offer flexibility, the car can also offer speed, at least in most areas. Public transport is just not flexible enough to be able to provide facilities for chained journeys.

Creating possibilities

Our societies create many chances and possibilities. Chances and possibilities are located somewhere: You should be able to reach them. Adolescents in the US, have, once their school bus has left, few possibilities of participating in anything without their parents' help.[2] A small action radius of mobility makes getting and fulfilling chances difficult. The unemployed without a car have difficulty reaching work on highway locations. Getting your driving license in the US feels like liberation and older people fear the moment they have to stop driving. In the Netherlands, this

---

2   See Bachiri, 2006, and Bachiri et al., 2008.

driving force is often overlooked, because the bike can be an alternative for a great number of journeys.

We will now turn to the description of the six driving forces from a more individual level.

Convenience and comfort

Modern people have, with growing prosperity, the chance to buy convenience. Growing prosperity also leads to a broader spectrum of consumption articles in households, articles that all have to be used, replaced, repaired, and carried. The car is convenient and comfortable for people and travelling with luggage is not very comfortable by other transport modes.

The wish for 'community light'

This driving force can be seen as the 'twin brother' of the geographical spreading-out of activities. People look for contacts in the vicinity of their house less often. Most family, and certainly most friends, live at greater distances. People like to meet people in their neighbourhood via the principles of 'community light'; they will identify with their neighbours in a way characterised by a certain distance in combination with easy moving contact. There is not much investment done in community light, so neighbours feel no need to watch other people's children. Playing in public spaces in the neighbourhood is considered, especially by the middle-class, with some anxiety in relation to driving children to the houses of their friends, where they can play in parents controlled environments. Also, visiting friends and family now means travelling greater distances than in the past.

The lack of ethical boundary

This driving force can be seen as the 'twin brother' of creating possibilities. As people see many possibilities and chances there seems to be no boundary. 'Everything that can be done, should be done', could be a motto in modern western risk societies. More people than ever will probably define their lives as a chain of events, spontaneous or self-created.[3] There are also many expectations; being mobile is such a customary practice that you are expected to drive to locations 100 km away the same evening for a joint activity. As we see a discourse on well-considered food, there is no start to define well-considered – ethical, thoughtful and responsible – mobility. With mobility, everything is taken for granted.

---

3 An interesting book on this theme is Schulze (1992, revised 2005), *Die Erlebnisgesellschaft.*

The car for identity

A modern western society has many abstract rules, standards and laws. These societies look ordered and people fit into the patterns created by these rules. However, especially in leisure time they also want to search for that identity, unhindered by rules and laws. Cars can be part of their identity. The consumption of the car can compensate for many unachieved expectations. Many people cherish their cars as the place to be, and the body to travel in.

Fears, feeling vulnerable and, and seeking protection

In risk societies people are aware of the risks related to their lifestyles. In the Second Modernity, risks are man-made and people try to mitigate and to avoid them. Confronting risks and anxieties with simple self-belief is not an option in our divided and spread out societies. Living individualised lives means standing alone in facing risks. Cars can protect people and households. Cars will be allies in the individual fight with the dangers of public spaces. Travelling by car makes you feel less vulnerable. Cars can be used as a defence mechanism.

The power of habit

The last of the ten driving forces applies to all ages. Living without habits seems impossible; habits arrange the complex worlds in a structured way. Most people stop thinking when something works. If shopping by car works the first time it is tried, the next time shopping is needed the owners will choose for the car again. Only with real and important changes, at 'tipping points', is there an opportunity for introducing new paradigms and for creating new habits.

This power of habits finalises the description of the ten driving forces of frequent car use. We have split these forces along the line; those which are more societal inspired – and those more inspired via individual motives. Other divisions are also possible. A useful division is for example 'Of the ten driving forces:

- Two are related to deliberate realising patterns and locations that are car oriented; spreading out of activities, highway locations;
- Three are broad working: creating possibilities, no ethical boundary, and habit;
- And at least five are social and cultural focused: the urge for flexibility, convenience and comfort, community light, identity creation, and feeling vulnerable and anxious.'

The situation that half of the driving forces are lifestyle-based makes discussing frequent car use rather difficult. You immediately enter a discussion of the lifestyles, norms and values of the discussants. Most scientists have moved out of this 'danger zone', or even do not come near. There is far more literature about changing spatial patterns, for example via building less car reliant cities, or about investing in alternatives for frequent car use via other transport modes. Research and literature on the social and cultural dimensions of car mobility, although growing, remains rather scarce!

4.4   From frequent car use to car dependence

Until now the focus in this book has been on frequent car use. Car dependence was introduced in Chapter 1. It is important to present a picture on frequent car use, the facts and figures, its cultural aspects and its driving forces, before moving to the specific compilation of journeys, locations, people and times that can be defined as car dependent.

Car dependence is a term that should be reserved for the lack of alternatives to make a journey in another way than by car. Or better stated, car dependence is the situation in which a journey can:

- be impossible
- or only with difficulty
- be made by another transport mode.

This description has three elements that need clarification. The first is the term 'situation'. There are five situations of car dependence:

1. Car-dependent locations; making a trip to these locations is impossible or difficult without a car. For example; reaching a highway location without any public transport.
2. There are car-dependent activities; undertaking an activity is impossible or difficult without the use of a car. For example; parasailing, or in general, luggage-oriented leisure activities.
3. There are car-dependent times; travelling at these times without a car is difficult, impossible, or even dangerous. For example; making a trip over longer distance at night.
4. There are car-dependent people; for them travelling without a car is impossible or difficult. For example; disabled people.
5. There are car-dependent societies; in some countries it is more difficult to live without a car than in others. Compare the United States with Denmark.

The second element to be clarified is the term 'impossible or difficult'. Here is room for debate. It is nearly always possible to give examples of making difficult trips without cars. When you have gone to a theatre in a city 25 km away from your home you could return after midnight by bike, or by walking and you can take a patient to the hospital on a delivery bicycle. However, most people will not see these alternatives to car use as reasonable. It is necessary to use, alongside 'impossible or difficult' the term 'reasonable'.

The last clarification to be made is the distinction between the factual lack of alternatives for car use, and the perceived lack of alternatives. In modern Western societies many households have grown alienated from the possibilities of public transport, which are just outside their scope. I will introduce in this chapter the distinction between factual car dependence (there is no other transport available to make the journey) and perceived car dependence (there are alternatives available, but people do not know).

4.5   Description of car dependence

There is relatively little literature on car dependence and the term is used in the public debate, but less in the research.

A short overview

We start with authors who define car dependence as synonymous with frequent car use. Take the description by Newman and Kennworthy, who published a standard book on Automobile Dependence in 1989. They stated a decade later 'automobile dependency is when a city or area of a city assumes automobile use as the dominant imperative in its decisions on transportation, infrastructure and land use' (Newman and Kennworthy, 1999, 334). And Litman (2002, 6) describes car dependence as 'high levels of per capita automobile travel, automobile oriented land use, and reduced transport alternatives'. However, by describing car dependence in this way the term 'dependence' is not clarified: 'when high levels of car use and car ownership are assumed to be indicative of a dependent relationship but without an explanation of the relationship between use and dependency, the equation amounts to a mere tautology' (Gorham, 2002, 108).

A better description is to be found in the first comprehensive study on car dependence, financed by the RAC (Royal Automobile Club of Great Britain) in 1995 and presented by a team assisted by Goodwin. In his *Mobility and Car Dependence* (1997) the car is seen as a creator of possibilities and chances. In the study, dependence is framed along two lines; the car as a drug, as addiction, and the car as parent, as a helper in societal traffic. Car-dependent journeys and car-dependent people are distinguished, and the situation that people know little of transport alternatives is identified. Goodwin concludes that only 15 per cent of the journeys have no alternative for car use, however 'for those trips with an alternative, actually using that alternative would generally mean about half the average door-to-door journey speed compared to the reported journey by car' (Goodwin, 1997, 457).

A decade later, in 2005, consultants Steer Davies Gleave published a report for Transport for London in which car dependence is framed as a lifestyle. In their view people are not dependent on the car, but on 'what it delivers in the context of the time constrained, dispersed and highly security aware lifestyles' (Steer Davies Gleave, 2005, 34).

Brindle approaches car dependence from the individual point of view in *Kicking the Habit: Some Musings of the Meaning of 'Car Dependence'* (2003). A car-dependent person in his view is:

a. somebody with such a lifestyle or such duties that only the car can provide the transport
b. somebody that cannot think of using another mode than the car.

And after this distinction he asks himself the question 'are we discussing lack of choice or a stubborn refusal to make the best choices when offered alternatives?' (Brindle, 2003, 64). He concludes that there is certainly also lack of choice; 'because of the way we have built our communities, most people will have no choice but to drive to shop, play or worship' (Brindle, 2003, 67). He identifies three possibilities for reducing car dependence; another spatial arrangement, a change in the mobility paradigms and, most important, a great change in value orientations.

Farrington, in *Car Dependence in Rural Scotland* (1998), illustrates the distinction between people who are dependent on their car because there are no real alternatives (this he calls structural car dependence, related to my factual car dependence), and people who have transport alternatives but put all their trust on the car (this he calls conscious car dependence, related to my perceived car dependence).

In *Travel Choice with No Alternative (2006)* Zhang presents a spectrum of car dependence in the literature. He also uses the distinction between factual and perceived car dependence. With perceived car dependence many people do not know the alternatives because their travel behaviour has become automatic.[4]

The border between factual car dependence and perceived car dependence cannot always be traced easily. On this issue, Lucas writes,[5]

> it is difficult to assess from the literature at what point people's behaviour at the individual level can be described as merely perceived reliance, or when this reliance becomes an actual dependence, or indeed, when it may be considered to be an effective dependency on, or addiction to the car. It is clear from the literature that what is being described is actually a spectrum of behaviour and a huge degree of subtlety is needed to be employed in determining whether an

---

4   Here we see a relation with the motive 'Habit'.

5   Lucas presented, at TRB 2009, a paper called 'Scoping study of actual and perceived car dependence and the likely implications for livelihoods, lifestyles and well-being of enforced reductions in car use'.

individual or household is genuinely car reliant or merely wedded to their car
because of habit, social norms or other non-physical factors. (Lucas, 2008, 7)

After this overview we can now prescribe the elements of a proper definition of car
dependence. Such a definition acknowledges:

a. the circumstance that car dependence can be about journeys and can be
   about lifestyles;
b. the recognition that car-dependent people, locations, activities, times and
   countries do all exist;
c. the need to specify the 'dependence' element in the definition;
d. the distinction between factual and perceived car dependence.

For element c. the dependence element's four phases can be identified. The
spectrum goes from using the car, via growing accustomed to the car, to relying on
the car and finally to becoming car-dependent.

Now we should operationalise the term 'reasonable'. What are reasonable
alternatives to the car and when is it reasonable to speak about car dependence?

We start with the bicycle. This transport mode is a real alternative up to ten
kilometres distance, provided that little or no luggage has to be transported and
provided that the weather is not all too bad. As a rule of thumb from statistics (see
Jeekel, 2011, 24) in the Netherlands for journeys up to 3.7 kilometres cycling and
walking dominate over car use. Over distances greater than 3.7 kilometres the car
dominates (also MON 2009, Table 7.5).[6] For the Netherlands, information can be
gathered from the CBS study *Verplaatsingsrepertoire Korte Ritten*. In this study
the two most popular transport modes for short journeys in the Netherlands are
compared. Cars are used for shorter trips with much luggage, in bad weather, in
the evening and at night, and for chained trips. The bike is used for daily shopping,
sports and for flexible short-distance trips.

Public transport has another ratio. For most journeys the time ratio over the
same distance is very unfavourable for public transport, relative to the car. With
the exception of distances over 50 kilometres the average journey on public
transport in the Netherlands takes double the time of the same journey made by
car. Above 50 kilometres this ratio diminishes to 1:1.7.[7] For these distances the
train can be an alternative. Although there are tracks where trains are faster than
cars – mostly to and from central city locations – there remains a difference in
travel time. However the train causes less stress, and train time can be spent in
more ways than car time.

---

6 Burbridge et.al (2005) found in the US a turning point at 5.8 kilometres, for more
active cyclists.

7 Data from; *Kennisinstituut voor Mobiliteitsbeleid* (2009c): *bij het scheiden van de
markt*; *vraagontwikkelingen in het personen - en goederenvervoer.*

Journeys of over 50 kilometres in the Netherlands account for 40 per cent of the car distance travelled, but only 12 per cent of the car trips (MON, 2009, Table 7.2). Trains, trams and buses rely on timetables and on defined routes. In contrast to walking and cycling public transport cannot reach all destinations in an area. It is impossible, certainly in the sparser populated areas, to travel at all times. Public transport is not geared for luggage.

To summarise; the reality of an alternative for car use diminishes when:

a. a distance between 10 (below this distance the bike) and 50 (above this distance the train) kilometres has to be travelled;[8]
b. there is no public transport stop in the vicinity;
c. much luggage has to be carried;
d. chain trips are to be made.

What are trips that reasonably need car use looking at the five 'situations'? Starting with *car-dependent locations*. On these locations service from public transport is minimal, and the connection with the car is maximal. Two types of locations can be seen. First the highway locations, for both working and living. Many working locations along highways have no public transport in their vicinity. Housing areas and estates mostly have public transport, but services are focussed on city centres, and we have already noticed that living near to highways means feeling rather 'footloose' about the destinations. Most easily reachable destinations by car cannot be reached by public transport. Secondly, rural areas and less densely populated areas are also car-dependent. Rural areas are, together with the peri-urban areas, the most car-oriented in Western Europe.

A number of *car-dependent activities* need the car as an instrument, mostly for transporting luggage and for reaching locations without any public transport. This is certainly the case for a whole spectrum of leisure activities. And it is the case for the weekly big shop.[9] Also the business to business travel along highways is difficult or not so easily made possible by public transport. Here, car dependence is seen as necessary and related to time constraints and time slots. Public transport is too slow for this type of travel.

At some moments in *time* there is no public transport available. The situation differs in different Western European countries, but it can be stated that in most countries later in the evening and the night, on Sunday mornings and often also in off-peak hours, public transport has little or no service. Walking and cycling are (especially at night and on late evenings) seen as rather dangerous activities.

What people can reasonably make no or little use of public transport? First the disabled: For them public transport use raises difficulties in getting on the bus or

---

8   Take into account that most commuting distances in the modern western societies are between 10 and 50 kilometres.

9   Note that in France more than 50 per cent of the big outlying shopping areas of Ile de France could not be reached by public transport (Orfueil, 2010).

train, but even more in being able to get somewhere relatively quickly. There are also, frequently, no designated helpers on buses, and at smaller railway stations.

Adults in families sometimes face situations where they have to combine trips rather quickly. The car is then used, although the bike could be an alternative, if there is no or little luggage involved. Only the highest level, most difficult, segment of escorting trips seems to be car-dependent, but often the car is chosen as the preferred mode, for convenience and the sake of comfort.

4.6   Car dependence quantified

In this section car dependence will be quantified. Quantification of car dependence is new, most authors present qualitative studies. However, from these qualitative studies it does not become clear how car dependent our societies are. Qualitative descriptions can never present figures on the magnitude of car dependence in western risk societies.

Quantifying car dependence is no easy job. Statistics are not framed for this theme. Quantifying car dependence means staying at the frontier between knowing and not-knowing. I will present results of three studies.

The first and the longest study was presented in my book on this subject (Jeekel, 2011, Chapter 5). I will present a methodology for quantifying car dependence. As the method is a first try, I will explain the different steps followed and the result will be an estimation.

The second study was published in a report called 'The Car in British Society', published in 2009 by the RAC Foundation and the last study is smaller and comes from Germany. I will present a classification of German cities on their car orientation.

A Dutch approach

The approach has five steps and partly uses material from three case studies (Jeekel, 2011, 307 and further). The results of the first study, on car dependence and children, were presented in Section 2.9. The results of the second study, on luggage and shopping with a focus on 'doing it yourself' and leisure, were presented in Section 2.9 and at the start of this chapter. And the results of the last case study, on highway locations were presented in Section 2.7.

First step: Identifying car-dependent locations
The three cases studies showed a number of factual car-dependent situations in the Netherlands and from these car-dependent locations can be identified. The situations are:

   a. the distance from home to school is greater than 3.7 kilometres and children are younger than 11 years. In these circumstances there is most probably –

in non-school bus countries! – no public transport available. The distance to cycle is too long. This situation will arise in the countryside.

b. a journey is made to a location at a distance of more than 10 kilometres from home, where there is no public transport service. The cycle journey is than too long. This mostly is the case in leisure areas – forests, lakes – and sometimes at peripheral shopping areas.

c. people living and working along the highway. It cannot be expected that people will than make voluntarily longer journeys from home to work.

d. reaching solitary highway locations, that are not or are minimally connected to the urban areas.

Car-dependent locations are:

1. the countryside, meaning rural areas, but also the low density areas nearer to built-up areas
2. areas with a lack of public transport services
3. highway locations.

Second step: Identifying car-dependent activities

From the case studies it could be concluded that car dependence is higher when activities involve more luggage to transport, and when these activities take place at times when there is little or no public transport available. The situations here are:

a. when a working parent transports children to school (with a start in the Netherlands at 8.30) and has to be at work at 9.00, while the work location is at a distance greater than 15 kilometres from the school. In these circumstances, which are rather common in the Netherlands, it is impossible to make the trip without a car.

b. escorting over greater distances where there is no or little public transport available.

c. a journey of greater distance is made for leisure with a start time, or an end time, or both, on which there is no public transport available, often the case with nature trips or when visiting friends.

d. where it is necessary to visit more highway locations on the same days, which is mostly the case in business-to-business activities.

e. activities where a great deal of luggage and/or purchases have to be transported. This is often the case with fun sports, camping, and with the weekly shop.

Car-dependent activities are:

1. escorting, in combination with work, shopping, volunteer aid, in tight time schedules;
2. visiting family and friends, at a greater distance, in the evenings;

3. many activities on Sunday mornings;
4. leisure activities with much luggage;
5. weekly shopping, and shopping with difficult items;
6. business-to-business activities.

Third step: Identifying car-dependent people
This is a combination step. The results of the first two steps are used to identify which groups of people are related to car-dependent locations and car-dependent activities. This is the list:

1. business-to-business workers, commercial services;
2. people that have to combine task in tight time schedules;
3. people living and working along the highways;
4. people living in the countryside;
5. commuters working on highway locations;
6. disabled and less able and persons needing care.[10]

For these categories the car will be so important that we can speak of dependence on the car.

A difference can be seen: For 1, 2 and 5 the car dependence is related partly to their activities. For 3, 4 and 6 the car dependence seems to be more general. We continue with this distinction in step 5.

Fourth step: Quantifying car-dependent people
This is the most difficult step. For each of the six categories of car-dependent people data can be found. The problems are related to the overlaps between the categories. For example; business-to-business workers can live in the countryside. They can be counted twice. The amount of overlap is not known. The status of the data presented can never be more than estimates.[11] The data are mostly from 2007.

1. Business-to-business people   To quantify this category the motive 'zakelijk bezoek in de werksfeer (business travel work)' of the MON (Mobiliteits Onderzoek Nederland) and the report Zicht op de Zakenrijder (Vereniging, 2007) are used. From this report it can be concluded that business to business people travel annually for their work 22.000 kilometres on average. In 2009 for this reason 12.54 billion kilometres were made (MON, 2009). This means that an estimate of 570,000 business to business workers are active in the Netherlands.

2. People combining tasks in tight time schedules   In the Netherlands there are 2.4 million family households. In 65 per cent of these households both partners work

---

10   A new category, relatively strong car-dependent. However, the car will often be a taxi or demand-driven public transport, mostly serviced by taxi companies.

11   I will work with band widths for some categories.

(SCP, 2008). So there are 1.6 million family households where at least one parent combines work and childcare activities. In many family households both parents have care tasks; 2 million people seems a reasonable estimate. It is estimated that 20 to 40 per cent of these households have to combine many activities in tight time schedules on a regular basis. This means that between 400,000 and 800,000 persons have a difficult job in combining tasks.[12]

For task combining, helping friends and family with health problems is also important. For this help (in Dutch *mantelzorg*, more than three months and more than eight hours care per week) we talk about 800,000 persons, 80 per cent aged 30–64, and more than half of this age spectrum 50–64 (data from SCP, *Mantelzorg in perspectief*, 2008). A fair share of the people giving volunteer aid have no work. An estimate can be that half of the volunteer aid helpers, 400,000 persons, are combining work and care. Estimating that 40 to 60 per cent of these persons have to combine many tasks in tight time schedules, leads to an estimate of between 160,000 and 240,000 people. This category has a band width of 560,000 to 1,040,000 persons.

3. People living and working along highways   In the case study on highway locations (Jeekel, 2011, 373 and further) I concluded that in 10 per cent of all Dutch households at least one member works and lives along a highway, whereby a distance of 1.8 kilometres from the nearest approach road is defined as the norm. In the Netherlands there are 7.2 million households, the average household having 2.2 people. This means that at least 330,000 people live and work along highways.

4. People living in the countryside   The statistics for persons living outside the more built up areas (cities but also villages) is rather outdated. From 2000 we have a figure of 925,000 inhabitants, with an estimated 450,000 adults. This means, again estimated[13] between 300,000 and 360,000 car-dependent persons.

5. Commuters to and from highway locations   In the case study on highway location I was able to conclude that 41 per cent of the employment in the Netherlands can now be found on a location in the vicinity of highways (Jeekel, 2011, 391). Twenty three per cent is located at a distance of 1 kilometre from approach roads, and another 18 per cent at a distance between 1,000 and 1,800 metres from these approach roads. The first 23 per cent can only be reached without a car with difficulty. I estimate that between 20 and 40 per cent of commuters has a reasonable alternative to car use available to these locations, meaning that 14 to 18 per cent will be car-dependent. For the 18 per cent of the commuters working at a

---

12   Note that I concentrate the more difficult task combining completely with the families. The truth will be more differentiated, certainly singles sometimes have problems combining tasks.

13   A part of the non-density areas in the Netherlands are to be found in the immediate vicinity of city outskirts and villages.

somewhat greater distance there are more alternatives available.[14] My estimation is that there will be alternatives for 50 to 65 per cent of the commuters, meaning that 6 to 9 per cent will be car-dependent.

At least between 20 and 27 per cent of the employed people in the Netherlands are car-dependent. Five and a half million persons are, daily, travelling to work (CBS Statline). Two thirds of these persons travel by car, and 20 to 27 per cent has a car-dependent work situation. This means between 740,000 and 1,000,000 persons.

6. The disabled, less able and persons needing care   Statistics for these groups are rather clear. We talk about 600,000 persons;[15] 500,000 are, in some form independent, half of them being older than 65 years.

To conclude this quantification Table 4.4 can be presented.

Table 4.4   Car-dependent adults in the Netherlands

Category of car-dependent adults	Numbers of persons
1. Business to business, commercial workers	570.000
2. People combining tasks in tight time schedules	560.000–1.040.000
3. People living and working along the highway	330.000
4. People living in the countryside	300.000–350.000
5. Commuters to and from highway locations	740.000–1.000.000
6. The disabled and the less able and persons needing care	600.000
Total car-dependent adults	3.100.000–3.900.000

There will certainly be some overlap between the categories, but I will count with these numbers. In the Netherlands there are 10.7 million adults that have a car at their disposal.[16] The car-dependent segment is then 29–36.2 per cent. For all Dutch adults we talk about a car dependence ratio between 23.7 and 29.8 per cent.

Fifth step: Quantifying car-dependent journeys
The number of car-dependent journeys is the sum of two elements:

---

14   Take into account that a distance of 1800 metres from an approach road in the densely populated Netherlands can mean that your working area is part of the urban fabric.

15   Here I follow SCP's *De ontwikkeling van de AWBZ – uitgaven* (2008). 586,000 persons had in 2007 some form of collective financed care.

16   N.B. This is not always the case, there are many households with two adults with driving licenses, sharing one car.

   a. the car-dependent journeys of the car-dependent persons

   b. the journeys for car-dependent activities of all persons (the activities from the Second Step)

We will concentrate on these two elements but first some difficulties. As can be seen the journeys by car-dependent people can, using this method, be counted twice. On the other hand the non-car-dependent persons make more car-dependent trips than just for car-dependent activities. The assumption I make is that these two journey types are more or less comparable in magnitude.

a. the car-dependent trips of the car-dependent persons   Car-dependent persons are not for all trips they make car dependent. A reasonable estimate can be made on the percentage of car-dependent trips they make. Stated earlier, that percentage is higher for the handicapped, persons living and working along highways and persons living in the countryside. That percentage is lowest for commuters to and from highway locations. Although important it is mostly only work-related trips that are car dependent for these persons.

Table 4.5   **Percentage of car-dependent trips of car-dependent people, estimates**

Category	% Car-dependent trips of all car trips
a. business to business, commercial	80%: this category drives much for work motives, and little for leisure
b. people combining tasks in tight time schedules	80%: many trips need a car, for flexibility reasons
c. people living and working along the highway	90%: not very much connection to the urban fabric, highway living
d. people living in the countryside	85%: distance to activities is the important criterion
e. commuters to and from highway locations	60%: trip to work is important, for other trips no car dependence
f. disabled and less able and persons needing care	95%: difficult without a car

Nearly 75 per cent of all journeys by the car-dependent people can be seen as really car dependent.

We estimate that car-dependent people make 10 per cent more journeys and travel 10 per cent more kilometres than the average for car drivers. This means that 29 to 39.8 per cent of all car journeys are made by car-dependent people. Of those journeys by car-dependent people 75 per cent is really car-dependent, meaning that car-dependent people bring in 22 to 30 per cent car-dependent journeys (of all car journeys), and 24 to 33 per cent of all car kilometres.

b. Car-dependent activities for all persons –
Escorting   In the Netherlands 23 per cent of all car journeys by family households are escorting trips, 32 per cent by mothers and 14 per cent by fathers (Jeekel, 2011, 327). The focus here is on the escorting journeys undertaken by non-car-dependent parents. As families make up 36 per cent of all households (30 per cent full families and 6 per cent single parent families) some 8.5 per cent of all journeys made are made for escorting reasons.[17] The bigger part of these journeys, an estimated 40 per cent, will be made by car-dependent persons. This leaves 5 per cent escorting trips for the non-car-dependent persons. My estimate is that 30 per cent of these trips can be seen as car dependent for them. This means an additional 1.5 per cent for escorting activities.

Visiting friends and family at greater distances and in the evenings   In the Netherlands journeys made for leisure are 38 per cent of the total. Thirty-nine per cent of these trips are made for this reason, this is 15 per cent of all journeys. And we know that 62 per cent of these journeys are made by car, this is 9 per cent of all journeys. A third of these journeys can be estimated as car-dependent, longer distance, on Sunday morning, in the evening. This means 3 per cent of the journeys can be defined as car-dependent, and around 4 per cent of the kilometres (longer journeys are more car dependent).

Culture, visiting pubs and restaurants   This motive accounts for 19 per cent of the leisure trips, is 7.5 per cent of all journeys. Most journeys are in the evening, and most journeys are on shorter distances. Fifty-five per cent of the journeys for this motive are by car, this is near to 4 per cent. An estimate is that somewhat more than half of these journeys will be car dependent; longer, in the evenings and the nights. This means 2.5 per cent of the trips and 2 per cent of the kilometres (most trips are rather short).

Activities on Sunday morning   On Sunday morning there is not much public transport. Alongside the visit to family and friends an estimate is that only 1 per cent of the total journeys and total distance is car dependent and made on Sunday mornings.

Leisure activities   Pure leisure motives are responsible for 19 per cent of the leisure journeys. This is 7.6 per cent of the total. Seventy-five per cent of these journeys are made by car. Journeys involving longer distances are more car-dependent. Somewhat more than 50 per cent are longer trips, longer than 10 kilometres. This means that 3 per cent of the journeys and 5 per cent of the kilometres are car-dependent.[18]

---

17   Compare this figure with escorting figures for Western European countries; 11 per cent on average, but Germany; 8.1 per cent. (see the Addendum).

18   This is without the trips made on holiday in foreign countries; the greater part of the camping and caravanning kilometres is not identifiable due to the organisation of the

Weekly shopping, and shopping with much luggage Shopping accounts for 20 per cent of all journeys. The weekly big shop accounts, as an estimate, for 25 per cent of these journeys, of which 80 per cent made by car, and are mostly car-dependent. Three and a half per cent of the journeys can be specified as car-dependent, and 3 per cent (most shopping trips are short,[19] these are the longer ones) of the kilometres.

Sixth Step: Estimating total car dependence

Table 4.6   Car-dependent activities in the Netherlands

Car-dependent activities	% car-dependent trips in the total of trips	% car-dependent distance in the total car distance
a. escorting	additional 1.5%	1.5%
b. visiting friends, family.	3%	4%
c. culture, cafe, restaurants	2.5%	2%
d. activities on Sunday morning	1%	1%
e. leisure	3%	5%
f. weekly shopping	3.5%	3%
Total	14.5%	16.5%

We can now summarise our findings. They give a first indication. The method can be used in the future, and should be improved.

Table 4.7   Summary of the results: Car dependence in the Netherlands

	% of all trips	% of all distance
a. car-dependent persons	22–30	24–33
b. car-dependent activities	14.5	16.5
In total	36.5–44.5	40.5–49.5

Dutch mobility statistics. Holiday travel outside the Netherlands is excluded in the mobility statistics.

19   Although in the Netherlands shopping accounts for 20 per cent of the trips, shopping accounts for only 9.5 per cent of the distance travelled.

This means that we can estimate that from all car journeys on average 40.5 per cent is car dependent, and that from all car distance travelled 45 per cent is car dependent.

A British approach

In 2009 the RAC Foundation for Motoring published a report *The Car in British Society*. Chapter 5 in this report addresses the issue; car use – a matter of choice or necessity (RAC Foundation, 2009, 110 and further). In this chapter the focus is on car dependence. At first the authors use the term 'car reliance'. They see a car-reliant trip as 'a trip where there is no other form of motorised transport available and the journey distance is too long to walk or cycle' (RAC Foundation, 2009, 116). They consider it difficult to assess when 'people's car use transcends into car "reliance", or when this reliance becomes "dependence" or, indeed, may be considered to be a pathological dependency or addiction' (RAC Foundation, 2009, 117). With a scheme the dynamic development of car dependence within a society over a period of time is clarified (RAC Foundation 2009, Figure 5.3)

And they present, in a comparable way as in this book, a scheme of trips, noticing three degrees; close proximity trips (A), trips made by car for which there are non-car alternatives (B), and trips that are car-dependent (C). In principle for B trips there is no car dependence, the car dependence is in the minds of people. The trips under C can be seen as the factual car-dependent trips. Within these C trips the authors see a division in structural constraints, related to societal factors (for example; no public transport available), and situational constraints, related to requirements of the trip makers or the trip itself, like having to carry luggage, or being disabled.

In the report some quantification is presented. In England the Ministry of Transport defined three Sustainable Travel Demonstration Towns; Darlington, Peterborough and Worcester. In 2004, Socialdata and Sustrans carried out a research on urban car journeys in these three towns, and found that respectively 56 per cent, 39 per cent and 46 per cent of the journeys made by car had alternatives,[20] leaving 44 per cent, 61 per cent and 54 per cent as car-dependent.

A little more than half of all journeys were seen as car-dependent, but there was a rather high differentiation. The set-up of towns, and probably the public transport services does matter, as we will see in the coming German example. In all three towns around a quarter of journeys were situationally constrained, the majority being structurally constrained.

The conclusion in the RAC Report is that half of urban journeys are now car dependent, but only a quarter of these trips is situationally car-dependent. For

---

20   Here also Socialdata and Sustrans had a somewhat similar approach as in this book. They had to define reasonable alternatives (RAC Foundation, 2009, 121) and choose for; walking 2 km, cycling 6 km and public transport; twice as long, with a maximum increase of 15 minutes travel time. Note that the research was done on urban car trips.

three quarters of the urban car journeys in principle alternatives could be created, however much investments in alternative transport modes will then be needed. This being said, the reality is that now *the car dependence rate in these towns is a little above 50 per cent.*

A German approach

Another approach to car dependence comes from Germany. In 2010 a paper was presented by Klinger, Kennworthy and Lanzendorf called *Mobility Cultures in urban areas; a comparative analysis of 44 German cities.* The authors looked at the relation between urban organisation and mobility choices. The central theme of their research was 'urban mobility culture', and this encompasses 'both material and symbolic elements of a transport system as part of a specific socio-cultural setting, which consists of mobility-related discourses and political strategies on the one hand and institutionalized travel patterns and the built environment on the other hand.' They noticed big differences in mobility organisation and mobility choices not only between cities but also between persons living in the same neighbourhoods.

The authors used a great number of indicators for this urban mobility culture. They choose: motorization (cars per 1,000 inhabitants, plus percentage of cars of more than 2,000 cc.), modal split, work trip distance and average speed of car trips, membership of cycling clubs, supply indicators on public transport and cycling possibilities, settlement density, share of 1 and 2 family houses, household income, share of elderly people and unemployment rates. They made, with data from 2002, a factor analysis.

In general the modal split on journeys differed quite a lot between the 44 cities. In general 48 per cent of all trips were made by car, but the range differed between 29 per cent and 65 per cent. Twelve per cent of the trips were made by public transport, with a range from 4 to 19 per cent, cycling accounted for 9 per cent of the trips, with a range from zero to 28 per cent, and walking accounted for 30 per cent of the trips with a range from 19–40 per cent.

The results of the factor analysis showed six groups of cities, with quite different rates on car dependence. At first there were the cycling cities, most university cities in the Northern part of Germany (for example; Aachen), cities with rather low car dependence. A second group were the transit metropolises; rather big cities with very well-equipped public transport systems (for example, Frankfurt and Cologne). A third group were smaller cities with an orientation to public transport, all situated in the former East Germany. These three groups of cities have relative low car dependence rates.

A fourth group consisted of 14 more average cities. And finally there are two groups of car-oriented cities identified; the car cities with a cycling potential, like Duisburg, and the car cities with a public transport potential, like Bochum or Wiesbaden. These cities have now high car dependence rates.

This German research identifies the importance of urban mobility cultures in realising higher or lower car dependence. Car dependence in its perceived form is created by individuals, but the supply of alternatives offered seems to be created by the broader mobility culture in a city or an area.

4.7 Car dependence in perspective

The driving forces for frequent car use were described in Section 4.4. Here I will look at the specific indicators for car dependence, related to these driving forces. The following indicators can be noted.

Table 4.8   Indicators for car dependence

Driving forces	Indicators for car dependence
Geographical spreading out of activities	Weekly shopping
The creation of highway locations	Living and working along highways Business to business mobility
The urge for flexibility, combining tasks in tight timeframes	Combining many tasks in tight time schedules
Creating possibilities for participating in life's experiences	Facilitating disabled, less able and persons needing care
Convenience and comfort	Bad weather
The wish for 'community light'	
The lack of an ethical boundary on mobility	
Cherishing the idea of identity also related to the car	Much luggage to be taken Visiting forests, nature, without public transport services
Fears, feeling vulnerable, seeking protection	
Habit in choice of transport modes	
	Activities at greater distance, in the evening or the night Activities on Sunday morning

In the Netherlands it is estimated that 40 per cent of the car journeys and 45 per cent of the car kilometres are car-dependent. In the three English towns we noted a somewhat higher car dependence, around 50 per cent. And in Germany we have no figures on car dependence, but can note a rather great difference between cities, with the car oriented cities probably higher, and the cycle and public transport oriented cities lower on car dependence. With these figures it has to be taken into account that car dependence is higher in less densely populated areas.

At least for the Netherlands it can be concluded that in the last 12 years a rather high growth in car dependence has occurred (Jeekel, 2011, 152). Fifteen years ago there were fewer highway locations, and these only existed for working, and the leisure pattern was less oriented on difficult to reach locations. Children were not escorted that much, and there was less combining of tasks. These developments have all been documented.[21] An estimate:

- for people living on the highway; 330,000 in 2007, compared with 30,000 in 1995
- for task combiners; 800,000 on average in 2007, compared with 500,000 in 1995
- commuters to and from highway locations; 870,000 in 2007 to 500,000 in 1995.

This could mean some 2.55 million car-dependent persons in 1995, compared to, on average, 3.5 million car-dependent persons in 2007. In 1995 9.6 million adults could use a car. This means that 26.5 per cent of the car journeys were made by car-dependent persons. With again 75 per cent of all these journeys being car-dependent journeys we arrive at 20 per cent car-dependent journeys by car-dependent people, and 22 per cent car kilometres.

On car activities it seems reasonable to reduce especially the rates on escorting and on leisure. I estimate a lowering to 11 per cent on journeys for car-dependent activities, and 12 per cent on kilometres.

Estimated, this means jointly 31 per cent on car-dependent journeys and 34 per cent on car-dependent kilometres in 1995. And this indicates a growth rate:

- in car dependence journeys (31 per cent to 40.5 per cent is 31 per cent; 12 years) is 2.6 per cent
- in car kilometres (34 per cent to 45 per cent, is 30 per cent; 12 years) is 2.5 per cent.

Unfortunately we have to work with estimates, but 2.6 per cent and 2.5 per cent are higher than the car mobility growth in the Netherlands, as we have seen on car kilometres 23.9 per cent in 14 years (1995–2009, see the Addendum) is 1.7 per cent yearly.

---

21   Highway locations can be seen by comparing maps, and on escorting facts were presented in 2.9 and see Jeekel (2011, 307 and further). The leisure pattern has changed, and is documented in Harms (2008), and data on task combining can be found in SCP (2006a, 22 and further).

Car dependence is growing faster than car mobility. Starting from 2007 to 2020 with the growth rates of the last 12 years in 2020 a car dependence of above 50 per cent will be reached, on journeys and on kilometres. Around 2030 around 65 per cent car dependence will be reached.

Here a comment should be made. With the focus on car dependence it remains clear that many persons and many activities are not car-dependent. To begin with, most students and schoolchildren. The energetic elderly and many commuters are not car-dependent, at least when they do not live in the countryside. Many persons carrying out escorting or helping friends and family with health problems are not car-dependent, because they have more flexible time schedules. And there are many adults that do not work. They also are not car-dependent.

Does this mean that all these persons are never dependent on a car to make a trip? Certainly not. As stated, there are more car-dependent journeys than car-dependent people. But also many trips are not car dependent; most shopping, mostly the school run, most visiting on daytime, and many commute routes are not car dependent.

At the end of this chapter we will return to the distinction between factual car dependence and perceived car dependence. Probably many people will see themselves or a number of trips as car dependent, but there are alternatives. They do not see these alternatives, or reject them immediately. There are four motives for this behaviour. First there is anxiety. People feel safe in their cars, and less safe on other transport modes. The car protects them better. Second is convenience, which here also includes flexibility. The car is just easier; you do not have to plan, as with public transport, or to give your own energy, as with walking or cycling. Especially with bad weather alternatives tend to be forgotten. The third motive is habit; people do not think about alternatives, they habitually take their cars. And the last motive is fun and freedom. The car offers the best journeys. Although there is an alternative, the car is nicer.

There is very little literature on perceived car dependence. In the *German Mobility Statistics of 2008* a question asked comes near (Abbildung 3.83, page 112). The preference shown is for the different transport modes for persons above 14 years. Six per cent is slightly mobile, 5 per cent uses, in the majority, the bike. There are 5 per cent public transport captives, unable to use another transport mode and there are 10 per cent *Stammkunden* (frequent travellers, almost always by public transport). Nineteen per cent of all persons are occasional public transport users. The majority, 55 per cent, favours the car. Thirty-six per cent of this last group is *Stammkunde* for the car, meaning they would hardly ever use another transport mode. Nineteen per cent are seen as car users, but also as potential users of public transport. However, at the moment they are normal car users.

So in Germany 55 per cent of all travellers prefer to use the car on their journeys. This is higher than the car dependence rates. I expected a relatively high perceived car dependence. A small questionnaire was drawn up and organised in summer 2010. We only asked four questions. These questions were:

- For how many years have you driven?
- For around 40 per cent of your journeys you depend on the car, and you cannot make the trip easy with another transport mode; do you consider this percentage high, low, or realistic?
- What are, for you, the most important reasons for taking the car?
- And why do you, when you know that there is an alternative, still choose the car?

One third of the 150 respondents were younger drivers (0–15 years driving experience), one third had driven between 15 and 30 years, and the last third for longer than 30 years. Forty per cent of the respondents thought 40 per cent car dependence realistic, 40 per cent considered that percentage too high, and only 20 per cent thought the car dependence higher. However, most respondents who answered that the percentage was too high regretted such a high percentage. A mixture of wish and fact seems to be the case here. However, these results show that people have a pretty rational view of car dependence, and perhaps even know that many of their journeys are in fact not car-dependent.

Respondents noted as the most important reasons for taking the car in the following order; luggage, bad weather, visiting highway locations, chain trips and task combining. Of lesser importance were; trips in the evening, escorting children, and trips on Sunday mornings. And even when there is an alternative, people choose the car out of convenience (70 per cent), for pleasure (15 per cent) and for safety reasons (10 per cent).

PART 2
Problems and Perspectives

In Chapters 5 and 6 I look at problems and challenges for car mobility and car dependence in society.

Chapter 5 is about fossil fuels and CO_2. Transport is now the fastest growing demand on fossil fuels and passenger car mobility is still almost completely dependent on them. But what will be the future of fossil fuels? What are realistic time schedules for moving to more sustainable car mobility via a more sustainable fleet? What influence will the objectives of policies against global warming have on our car-dependent societies?

Chapter 6 is about people. In modern western societies some 20 per cent of households do not drive or do not own cars. Who are these carless households? And how do they cope with living without a car? In this chapter attention will also be paid to the price of driving, especially for poorer households, and to aspects of social exclusion related to transport.

In Chapter 7 I will look at four possible scenarios for car mobility. What can be expected? On the one hand we have frequent car use (with mostly still some growth), car dependence (growing relatively fast), on the other hand we see problems for poorer households, for specific carless households, and especially problems with fossil fuels and with CO_2 emissions.

Finally, in Chapter 8 the focus is on the social and cultural aspects of car mobility in our society. We return to the description of the risk society from Chapter 2 and will look at relevant lifestyles aspects in modern western risk societies, in relation to car mobility. The focus in this chapter is on further research.

Chapter 5

Frequent Car Use: Energy and Sustainability Questions

5.1 Introduction

This is the first of the two 'problem' chapters. The focus in this chapter is on the challenges that the global energy situation and the global warming present to frequent car use. Central elements in this chapter are the future of fossil fuels and the need to reduce CO_2 emissions substantially. Both elements will create new frameworks for car mobility.

In Sections 5.2 to 5.4 the focus will be on energy. Fossil fuels are still essential for car use but the situation regarding these fossil fuels is not stable. Situations can arise of excessive demands in relation to the supply of fossil fuel. In Section 5.2 the current situation and an initial look in the future are presented. Section 5.3 looks at vulnerabilities from the perspective of both car users and wider societies relating to higher fuel prices and in Section 5.4 the global issues relating to the delivery of fossil fuels over the coming decades are central. We may face temporary or permanent delivery problems in situations where alternative fuels for the car are still in their early stages of development.

In Sections 5.5 to 5.8 the focus is on global warming and CO_2. In Section 5.5 the current situation relating to car mobility is presented. Mobility is responsible for the only still growing volume of CO_2 emissions. In Section 5.6 the objectives regarding CO_2 emissions and combating global warming are described, both for society at large and the transport sector. In the solutions for transport two themes are central; energy efficiency and vehicle technology. These themes will be introduced in Section 5.7. And in Section 5.8 we will look at the perspectives on change, showing the responsibilities and problems faced by consumers, producers and decision makers.

Sections 5.9 and 5.10 focus on integration. Reaching CO_2 objectives in times of fuel scarcity brings new perspectives on frequent car use and car dependence. Can a shift towards new forms of car mobility be made without causing crises? Scenarios will be presented, related to earlier paragraphs and in Section 5.10 some scenario exercises in other countries will be introduced. The analysis clarifies the need to discuss volumes of car traffic, car travel behaviour and the introduction of a societal package of measures, adjacent to the technical measures discussed in Section 5.7.

5.2   Fossil fuels, 2005–2030

Car use is oil dependent. Ninety-seven per cent of the fuel used in transport is fossil fuel. The fate of the transport sector and car use as an important element in society is, for at least the next 20 years, related to the fate of oil in our world. In this section the focus will be on the demand for oil, the supply, the discrepancies between these two, and the behaviour of consumers, enterprises and governments.

Demand for oil

Until recently a strong growth in oil demand was foreseen. In 2002 this demand was about 77 million barrels per day, in 2005 a little below 84 million barrels a day. The expectation was a growth in demand to 99 million barrels per day in 2015. The increase in oil consumption was foreseen in 2007 by the International Energy Agency (IEA) in *World Energy Outlook* at 1.4 per cent increase yearly resulting in a growth to 115 million barrels per day in 2030. This 1.4 per cent increase corresponded with a BNP growth in the world of 3 per cent (globally 2 per cent in the OECD countries, and 5 per cent in the developing world).

In 2008, the IEA introduced, looking at the world economic crisis, a new prognosis with lower yearly increases and the IEA now expects a yearly growth of the oil demand between 0.4 per cent and 1.4 per cent yearly. On average the oil demand for 2015 will be a mere 90 million barrels per day.

The oil situation can be defined as turbulent. Prices fluctuate strongly, growth figures, on demand, but also on supply, change by the month. A general expectation is however that, accepting greater fluctuations, fuel prices will be rather high the next decade, and will probably rise and there will be growth in the demand for oil.

There are three greater determinants on this demand growth. The first is the significant growth of the economies of the non-OECD countries. The greatest economic growth rates are expected in China, Brazil, the other Asian economies and the oil producing countries of the Middle East. China is beyond concurrence and is seen as the new factory of the world. The demands for fossil fuels will rise enormously (see Canzler, Knie and Marz, 2006; Schipper and Ng, 2004; Weider 2004; Zhu, 2005).

This growth creates a shift in the spectrum of oil demanding economies. At this moment 70 per cent of the oil demand for transport comes from the OECD countries. This share will probably diminish to 55 per cent in 2030 and 45 per cent in 2050 (IEA, 2007).

A second determinant is the growth in car ownership and car use in the whole world. A core article on this subject is *Vehicle Ownership and Income Growth, Worldwide 1960–2030* (Dargay, Gately and Sommer, 2006). On the basis of an advanced model they calculate that in OECD countries the growth in car ownership will be relatively modest, around 0.6 per cent yearly from 2007 for

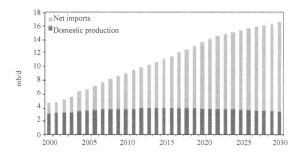

Figure 5.1   China's oil balance
Source: IEA (2007b). World Energy Outlook.

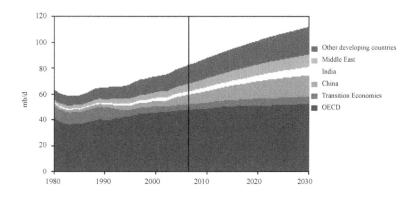

Figure 5.2   Primary oil demand, parts of the world
Source: IEA (2007b). World Energy Outlook.

the period until 2030.[1] There is already some saturation to be seen. In the non-OECD world other growth figures are expected, around 3.5 per cent growth in car ownership yearly. It is expected that China will have 270 cars per 1,000 inhabitants in 2030, comparable with Western European countries in the early 1970s. The total number of cars will, in their opinion, rise from 850 million in 2007 to 2 billion in 2030.

World population will probably grow from 6.2 billion inhabitants (2002) to 8.4 billion in 2030. In particular, the richer economies in the developing world will yearly show a rise in car use and in car ownership. Cars are being purchased when the income of a household reaches a threshold somewhere between 10,000 and 20,000 dollars.

---

1   We note in Chapter 2 that car ownership is still growing relatively fast in our selected countries. There is indeed a slowing down of the increase.

Shafer and Victor argue in *The future mobility of the world population* (2000) that in 2050 an average inhabitant of our world will travel as much as a Western European did in 1990.[2] They indicate that an average Indian will, in 2050, have the mobility pattern of a Western European in the early 1970s. They foresee a mobility growth, factor 5.

The last determinant is the growth in international trade. Global trade means high extra amounts of freight traffic. Transport is now responsible for nearly 60 per cent of the increase of oil demand. In the longer term, until 2050 the World Business Council for Sustainable Development expects, in their study *Mobility 2030* (2004), growth with 2005 as a basic year:

- a tripling of freight kilometres
- and a rise in passenger traffic from 30 trillion kilometres to 70 trillion kilometres.

The transport sector is raising its energy efficiency levels. This means that in the vision of the World Business Council the demand for energy will only double between 2005 and 2050.

Concluding the increase in demand for oil, coming from transport, will be 40 per cent higher in 2030, compared to 2008.[3] Transport uses now about a third of all oil. This share will increase to nearly 43 per cent.

Supply of oil

Where there is a reasonable amount of consensus on the range of the demand of oil, such a consensus is lacking on the supply side. On this side two pairs of terms are particularly important. The first pair is conventional oil against non-conventional oil. Conventional oil can be extracted fairly readily. Non-conventional oil (for example oil from the Canadian tar sands) is more difficult to extract and demands greater investments. It is difficult to make an inventory on the amount of exploitable non-conventional oil. The other relevant pair of terms is the physically possible supply as opposed to the factual supply. Geologists focus on the physically possible supply. However this supply is not what reaches the market. The factual supply reaches the markets, this is the supply that the producers, the governments of the oil states and the oil companies present. There are numerous reasons for limiting the physically possible supply.

---

2   The starting situation on mobility in developing countries is described in Freund and Martin (1999); *Driving South: The Globalization of Auto Consumption and its Social Organisation of Space*.

3   Note that the increase in the total oil demand until 2030 was in the EIA prognosis 2007 around 35 per cent (from 84 million to 115 million), and in the EIA prognosis end 2008 between 20 per cent and 35 per cent. Compared to other societal sectors Transport will be rather oil consuming in the coming decades!

There is also debate about the possible physical supply. In general there is a difference of opinion between experts working from peak oil theories and experts challenging these theories. 'Peak oil' can be defined as the moment at which oil supplies will no longer increase. Or more explicitly – peak oil is about the maximum volume of oil production in a region. The basic question about peak oil is not 'when will there be no oil left?' but, 'when can an increase in oil production no longer be realised?'

The term 'peak oil' was coined in the 1950s by the American geologist Hubbert. With his theory he was able to predict the peak oil moment in American oil extraction. Many peak oil authors[4] now see the world as their study area and think that the world peak oil moment will take place before 2020. They focus on established reserves and reserves yet to be established, but there are many uncertainties about the magnitude of oil fields in the Middle East and in Russia.

There is, however, consensus between oil experts on reaching peak oil in the non-OPEC countries in a few years from now. The differences are on the far more important OPEC reserves and on the prospects of winning non-conventional oil[5] over the next decade.

Hirsch et al. present, for the United States Federal Government, a balanced vision under the title *Peaking of World Oil Production: Impacts, Mitigation, Risk Management* (2005). They start with some logic 'when the world oil production peaks, there will still be large reserves remaining. Peaking means that the rate of oil production cannot increase' (Hirsch et al., 2005, 12). They take authors opting for early peak moments seriously because they identify minor results from the recent exploration activities of oil companies. They also clarify the difference between geologists and economists. Geologists are mostly pessimistic about the technical perspectives of reservoirs[6] while economists use other arguments. They suspect that OPEC countries are playing an information game' and are convinced that higher oil prices will lead to disclosure of new possibilities. But Hirsch et al. observe on non-conventional oil; 'we know of no comprehensive analysis of how fast the heavy oil production might be accelerated in a world short of conventional oil' (Hirsch et al., 2005, 6).[7]

---

4   To name, Aleklett (2008), Campbell (n.d.), Goodstein (2007), Laherrere (2004), Leggett (2006) and Simmons (2008).

5   An illustration is offered by Aleklett, the chairman of the international Peak Oil organisation. He mentions the prognosis from the World Energy Outlook 2004; the production of conventional oil will not peak before 2030 when the right investments are being done, especially in the OPEC countries. The IEA bases its vision on the estimates of the US Geological Survey, but when these estimates are too high, or when investments are not sufficiently made, peak oil can be with us 15 years earlier, around 2015.

6   Many writers on peak oil are geologists with former careers in the oil industry.

7   Hirsch's 'vision is broad. That is not the situation with most peak 'oilers' and their opponents. For example Lynch dislikes the peak oil approach. In *The New Pessimism about Petroleum Resources: Debunking the Hubbert Model and Hubbert Modellers* (2004), he argues that their vision is too static. Peak oilers see economic and political

In recent times writers on peak oil have acknowledged that after the peak there will not be a sharp descent, but rather a long plateau; the amount of winnable oil will remain for longer time at the peak level.

Two broader visions are bringing the debate into wider perspective. In *Running Out of and Into Oil* by Greene et al. (2003) the view is defended that if the conventional oil production will not peak before 2020, the growth in oil supply will diminish soon after this date. The authors are sceptical about pro-active signalling 'neither increasing costs nor falling proved reserves may provide a timely signal that a turning point is immanent' (Greene et al., 2003, 55–6). In *The Debate over Hubbert's Peak: A Review*, Al-Husseini (2006) paints three possible scenarios for oil supply. In his middle scenario, the peak oil moment is foreseen around 2020 with a production between 90 and 100 million barrels a day.

To conclude, what can be expected is a peak moment within 15 years for the production of conventional oil. This moment can shift by one or two years. What is not known is the volume of oil expected which can be expected from non-conventional sources, but the expectations are relatively low.

The political and economic situation around oil

More important than the physical boundaries are the political and economic boundaries around oil, defined by the question; how much of the existing supply is brought to the market by the oil producers? Answering this question creates a subtle game between matters of will and matters of possibility. In *Turbulence in the next decade; an essay on high prices in a supply-constrained world* (2008), a Dutch report from the Energy Programme of the Institute for Foreign Affairs (Clingendael), Jesse and van der Linde start by explaining that OPEC has decided to decrease investments in medium priced oil (regular oil). With these reduced investments about 100 million barrels a day can be reached in 2030 and not the 115 million that might be necessary.[8] In their view a gap between supply and demand will have huge great effects. When slightly more than half of the demand can be provided by a supply the world will reach a situation of scarcity of fossil fuels.

The question is; why are the OPEC countries acting this way? Central in their analysis is the situation that OPEC, Russia and Mexico are keeping a part of their oil supply, about 8 to 10 million barrels a day, off the market. The investments needed to win this oil are minimal. If the oil producing countries would bring this oil onto the market, prices would rise more slowly, demand-supply relations would be better, and political turbulence would lessen.

---

effects as geological restrictions, which means missing the causalities between exploration, discoveries and production of oil.

8   Jesse and van der Linde take as a basis the EIA 2007 prognosis, lowered, as we know in 2008. The gap between demand and supply will in this recent prognosis be lower.

There are important reasons for this behaviour by the non-OECD oil producers. They have noticed that due to fighting global warming and high oil prices the OECD countries are all investing in alternative fuels and in non-fossil powertrains for cars and trucks. They consider that oil demand will be less secure in this decade as a result of these investments. The following observation seems essential: 'Given the fact that the OPEC countries feel that they cannot control the transition process towards an energy sustainable world they are worried that one day their investments will become obsolete' (Jesse and van der Linde, 2008, 66). Most OPEC countries are not investing heavily in new explorations, which will put less oil on the markets in 2030 than will be available, thus creating a period in which transport in particular will still be strongly fossil fuel dependent while scarcity, along with high oil prices, will create high revenues for oil-producing countries.[9, 10]

Alongside this scarcity will be a distribution problem. Who will get the most of the scarcer fossil fuels? The developing world, with China as a landmark, or the established OECD countries? Also there is global dependence and interdependence; it is clear that there will be more fossil fuel for the economic take off of the developing world when the OECD countries succeed in reaching high levels of energy efficiency over the next decades.

The behaviour of the OPEC countries will have two effects; oil prices will fluctuate, at a high level, with an occasional steep drop, and there will be demands that cannot be met. The necessary fossil fuel is not being delivered, and what the reaction of the demanding parties and countries will be is not known. On this theme Jesse and van der Linde '*without strong global leadership, tight commodity markets and resultant high prices could easily drive us towards an ugly world, with nations engaging in destructive competition for scarce energy resources*' (Jesse and van der Linde, 2008, 29).[11]

---

9 With less nuance but in the same direction Adelman, the oil expert from MIT says in *The Real Oil Problem* (2004) 'Beginning 1999 and with the half-hearted cooperation of Russia, Mexico and Norway, OPEC was able to constrain world production and thus raise prices', and 'OPEC's constant concern has been to restrict supply and resist downward price pressure' (Adelman, 2004, 12).

10 The OPEC countries could have shared their foreseen risk with the oil companies. However, in most OPEC countries nationalism prevails. Oil production is now for 80 per cent of OPEC supply in the hands of state owned companies. Western oil companies get only concessions in the more difficult winning areas, and have 15 per cent of the reserves. In the non-OECD world the capital investments of oil companies go to the difficult location (Jesse and van der Linde, 2008, but also Bruggink, 2006).

11 A possible win–win strategy may be to obtain long term commitments from OECD countries to purchase fossil fuels from the OPEC countries, in exchange for investments from OPEC countries in mid-term oil, while possible with wider market entrance of the oil companies. It is questionable whether such a general strategy is possible in our world. More mutual trust is needed.

5.3   The price of fossil fuels and its impact on car drivers

The global demand–supply situation will give rise to two impacts. The first impact is on oil and petrol prices. This impact will be discussed in this section and the second impact will be on uncertainties in the delivery of fossil fuels. This impact will be discussed in Section 5.4.

Consensus is growing about changes to oil prices. It is expected that prices will fluctuate more and will, on average, become higher. There will be fast reactions to small changes in supply, and on temporary loss of demand, due to the economic problems in demanding countries and areas. In this paragraph we will look at the elasticity in oil prices, at changes in elasticity through the years and at the way car drivers adapt and adjust to higher prices for petrol and diesel. This section will end with a description of adjustment questions in areas that are most vulnerable for higher petrol prices, the suburban areas of Australia and Northern America. In these suburbs one can see through a magnifying glass what will happen when petrol prices will rise.

In *The Drivers of Oil Prices* (2007) Fattouh looks for the reasons behind high oil prices and introduces three approaches. The first is the economy of the exhaustion of natural resources. Oil can be exhausted, and most analysts conclude that oil prices will rise. In Fattouh's opinion it is questionable whether oil should be treated that way, because it is probably not exhaustion that is the problem, but the shortage of investments in exploration, and the way these investment costs are charged. The next approach is the demand-supply chain. The demand for oil is inelastic. The demand side in this approach is about incomes and prices, and the supply side is about prices, reserves and the behaviour of OPEC. The framework is too loose to predict prices in real life. Fattouh's last approach is an informal approach. He makes an analysis comparable to Jesse and van der Linde. The oil market on the OPEC side seems to have lost its spare capacity. A high demand leads then to high prices. It is unclear whether the disappearance of this spare capacity is a result of unwillingness to invest in oil infrastructures in the OPEC countries. Also unclear is the role of speculators in the price forming process. Fattouh concludes that the three approaches by themselves do not clarify much, but together some clarification arises. He argues for unilateral explanation modelling.

How do car drivers react when confronted with fluctuating, high prices for petrol and diesel? To answering this question a view of the literature on the elasticity of changes in fuel prices will be useful. There is a huge variety of studies, research methods and research strategies. At first sight empirical studies show great differences. More elaborate analysis however shows more consensus.

Brons, Nijkamp, Pels and Rietveld published *A Meta-analysis of the Price Elasticity of Gasoline Demand: A System of Equations Approach* in 2006. Their result is a price elasticity of –0.53, showing that car drivers are not very sensitive to changes in price of petrol. They observe that the price sensitivity of drivers is higher in the longer term than in the shorter term, because a longer period gives car drivers more opportunities to change their behaviour. Making changes in fuel efficiency or changes in car ownership (buying a more economic and efficient car, getting rid of the second car), are long term changes. In the shorter term the change is in the vehicle kilometres travelled. Here only minimal changes can be noted and the price sensitivities in the US, Canada and Australia seem to be smaller than in the other OECD countries.

A study by Hughes, Knittel and Sperling (2006) comes to comparable results. In *Evidence of a Shift in the Short Run Price Elasticity of Gasoline Demand* they conclude that Americans have become less sensitive to changes in fuel prices. They explain this result by pointing at the dependence of many Americans on their cars and on the situation that with growing prosperity fuel now has a far smaller share in household consumption. They mention the need to travel greater distances, with no possibility of alternative transport.

In *The Demand for Automobile Fuel: A Survey of Elasticities* by Graham and Glaister (2002), a difference is made between price elasticity in the short run and price elasticity in the longer term. They also conclude that price elasticity in the short term (–0.2 to –0.3) is far lower than in the longer term (–0.6 to –0.8). They conclude, 'Therefore it may be right to say 'it won't make much of a difference' or 'people will use their cars just the same', but only in the short term. The evidence is clear – and remarkably consistent over a wide range of studies in many countries – that in the long term there is a significant response, albeit a less than proportionate one' (Graham and Glaister, 2002, 20). The changes in the longer term find their rationale in growing awareness of car drivers, especially on improvement of their fuel efficiency.

Litman approaches the same theme from another angle in *Appropriate Response to Rising Fuel Prices* (2008). He shows that countries with higher fuel prices did realise far better energy efficiencies in their cars, than countries with low fuel prices. The more fuel efficient countries have fewer kilometres travelled per person, more travel alternatives, relevant housing and working locations in the vicinities of their homes and have, thus, a lower household budget for transport. He concludes, 'rather than trying to minimize fuel prices it is better to allow prices to rise and to help consumers, businesses and communities reduce total fuel costs by increasing vehicle and transport system efficiency' (Litman, 2008, 15).

In line with Litman is the Shell publication *Signals and Signposts,* an update of their *Energy Scenarios to 2050,* which will be introduced later. Shell signals (Shell, 2011, 40) that the US shows an average of double the distance travelled by European drivers annually. Shell analysed, for 20 developed economies, the determinants of the difference in distance travelled and noted two important factors. The first are decisions which lead to the development of highly dispersed urban infrastructure or city sprawl; this accounts for 60 per cent of the kilometres travelled.[12] The second determinant is life style choices that tend to increase the

---

12   Here a cross link can be made with the German study on different German cities identified on their mobility culture. Also, here the way in which the cities were built was significant.

propensity to drive, and these account for 40 per cent of the kilometres travelled. The differences in population density do not explain very much.

In the Netherlands the *Kennisinstituut voor Mobiliteitsbeleid* (KiM) published a short study on oil prices, economic growth and mobility (*Olieprijzen, economische groei en mobiliteit*, 2008a). The researchers introduced a 17 per cent higher oil price in 2020. Using an advanced model this rise resulted in 4 per cent less mobility. Two and a half per cent (somewhat more than 60 per cent of the change) finds its reason in fewer kilometres travelled. The car is used less for shopping and social leisure motives (–6 per cent). Business travel is not very sensitive to rising prices (–0.6 per cent) and commuting is on an average level (–2.5 per cent). The other 1.5 per cent (somewhat less than 40 per cent of the change) is related to diminishing car ownership. In the areas outside the Randstad this 4 per cent less mobility leads to 13.5 per cent less congestion. In the Randstad the impact is lower, around 11 per cent less congestion. Here is a great latent demand!

In a later publication *Verkenning Autoverkeer 2012* (2008c) the KiM observes that lower fuel prices could lead to a higher than expected growth in car mobility. A 4 to 11 per cent lower fuel price would, according to their modelling, lead to a growth of 11 to 14 per cent on car mobility in the years 2007–2012.

One conclusion is that with rising fuel prices the effects in the long term will be greater than in the short run. People will drive less, especially fewer kilometres, for shopping and leisure and they will start thinking about changes in their driving behaviour, car choice, car ownership, sometimes in locational choices, but certainly in fuel efficiency.

From these elasticity studies a first series of behavioural changes arises, but the spectrum is broader. This can be shown when a second series of studies is considered. These studies, primarily made by social scientists from Australia and the United States, show in detail the social consequences of higher fuel pricing. In these OECD countries (Canada also included) the rise of fuel prices has greater effects than in Europe. There are at least two reasons: The first is that the price of oil is a substantially greater element in the prices of petrol and diesel in these countries. They avoid the disturbing impact of high taxes on petrol and diesel. In times of rising fuel prices this tax component has a mitigating effect. The second is that these countries, having vast suburban areas, are more car-dependent than most European OECD countries.

The most extensive literature on the effects of high prices for fossil fuels comes from Australia.[13] Important in this respect is the Urban Policy Programme of the Griffith University in Brisbane. In *Shocking the Suburbs: Oil Vulnerability in the Australian City* (2005) Dodson and Sipe start by observing that a situation of 'uneasy oil' has, in parts of Australia, more impact than just higher prices. The cities of Australia have rich city centres and a ring of richer suburbs immediately around them. These areas have rather well-developed public transport which is,

---

13   On the relation between fossil fuels, CO_2, transport and society at large, see Glazebrook and Newman (2008).

however, missing in the so-called outer suburbs. It is in these poorer suburbs that the lower income households live, often with mortgage problems. The distance from these outer suburbs to health care and jobs is quite long, and this distance can only be overcome by the car. Rising fuel prices have a huge influence on household budgets in these outer suburbs. Then there is the so-called 'transport stress'; households with incomes 40 per cent lower than the Australian average spending more than 20 per cent of their low incomes on transport. Australian politics has only recently picked up these problems of lack of public transport and high rising fuel prices. Dodson and Sipe (2005, 25) 'the neglect of the outer suburbs may prove a strategic miscalculation'.

Dodson and Sipe (2006) also notice that in some areas high fuel prices give rise to changes in financial and consumption patterns. Oil prices have an influence on inflation levels in the Australian economy. Some taxes have been cut, on the basis of a rationale set out by former Prime Minister Howard 'the high price of petrol is having a depressing effect on people's livelihoods, people's incomes'. Dodson and Sipe consider that the problems of the outer suburbs could be mitigated by creating a public transport system, at the costs of spending money on highways.

In *Addressing Oil Vulnerability through Travel Behaviour Change* (Meiklejohn, 2008) the author presents the results of the adjustments made by Australian car drivers as a result of the high petrol prices of 2007. Sixty one per cent use their cars less, 59 per cent try to combine trips, 29 per cent saves money on other spending (with going to a pub, luxurious food, and newspapers often mentioned) and 19 per cent makes more use of public transport. The use of city buses has risen.[14]

In the United States the same trends can be seen. Cortright shows in *Driven to the Brink: How the gas price spike popped the housing bubble and devaluated the suburbs (*2008) that house prices fell the most in price in urban areas where the central city was no longer vital, such as Miami, Phoenix or Detroit. Where the core cities were vital – like Chicago, Seattle or Boston – house prices decreased less. Housing prices in the suburban fringes, in the exurbs, unilaterally had the highest decreases in value. With high fuel prices these areas also have the highest transport costs. A certain return to the core areas of the vital cities seems to be taking place; 'While it may be premature to predict that the nation's suburbs will become ghost towns or slums, it seems clear that the trend towards ever increasing sprawl is ebbing'(Cortright, 2008, 21).

In *Driven to Spend: Pumping dollars out of households and communities* (2005), the Surface Transportation Policy Project stated that households in regions that invested in public transport now have the advantage; they have reasonably-priced, alternative options for transport. Households with lower incomes in regions where no investments in public transport has taken place have difficulties and households in better-endowed regions can save money that could be spend on creating a really good intermodal transportation system.

---

14   Since 2006 in the US, the vehicle miles travelled per person is diminishing, from 27.6 miles per day in 2002 to 27.2 miles per day in 2006 (Polzin, 2006).

It is clear that problems in Western Europe are on a smaller scale. Fuel prices did not fluctuate as much as in Australia or in the United States. Most urban regions have transport alternatives, alongside cars. However, smaller variants of the described patterns can be found in Europe's rural and peri-urban areas, as was seen in Chapter 2. Those making long commute journeys every day that are only partly paid for by their employers will start thinking about alternatives. In *Road Pricing: a transport geographical perspective* (2007) Tillema shows that some shopping and social leisure traffic will disappear. Looking more specifically at vulnerable groups it can be expected that poorer households with intensive car mobility can fall into a 'danger zone'. In the Motivaction classification, risks are highest for the Social Climbers. It is expected that is this mentality group many households can be found who press for lower taxes on petrol and diesel.

At this time there is little social science-oriented literature on the consequences of peak oil on living in modern Western risk societies. One exception *is Children and Peak Oil: an opportunity in crisis* (2007) by Tranter and Sharpe. They relate to a bigger societal story; 'Cheap oil has enabled many parents in developed western societies to engage their children in lifestyles that are aimed at preparing them for the best possible adulthood in terms of a consumerist lifestyles ... however, with the end of cheap oil may come significant changes in the ways in which we treat and conceptualise children' (Tranter and Sharpe, 2007, 183).[15]

5.4   The certainty of delivery around oil

Thus far the adjustments do not look dramatic but this changes when we look at the impacts on a higher scale; too little supply for the worldwide demand. From 2015 there will be the challenge of oil scarcity, with virtually no alternatives for car mobility ready. In 2008 the Dutch Minister for Economic Affairs and Energy published *Energierapport 2008*. This report states that there are enough resources worldwide to meet energy demands but it is recognised that the spread of oil resources is limited, while most sources are found in unstable and unfriendly countries. The lagging behind of investments on extraction is seen as a serious problem: 'It is difficult to say whether the crucial investments will be done on time' (Ministry of Economic Affairs, 2008, 32). And the sentence 'The question whether enough will be invested in the energy producing countries is further complicated by the fact that these countries want demand security, before they invest. The effort of the European Union on energy efficiency, energy saving and sustainable energy is seen as a factor of influence on this demand security by the oil producing countries' is crucial (Ministry of Economic Affairs, 2008, 33). The report also mentions that 'In the near future interruption of delivery of oil and

---

15   Also interesting is their sentence, 'after peak oil, few 5-year-old children will be driven across cities on weekends so that they can play soccer with other 5-year-olds' (Tranter and Sharpe, 2007, 192).

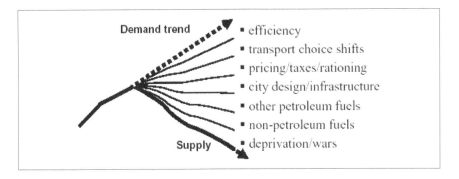

Figure 5.3   Bridging the gap between supply and demand of oil
Source: Robinson and Mayo (2008).

gas or great fluctuations of prices for these commodities cannot be impossible' (Ministry of Economic Affairs, 2008, 33).

Until 2030 the world should be busy with the first phase in the battle against global warming. This means working on less dependence on fossil fuels. Transport can reach its global goals only by investing in alternatives for fossil fuels, and in energy efficiency at a greater scale.[16]

This transition to sustainable energy for transport and car use will take more than two decades. To quote our Clingendael authors: 'any change, any replacement of oil by an alternative fuel will take a long time before it can make any realistic impact' (Jesse and van der Linde, 2008, 69). The interesting question will be how the different time schedules of:

- the development of new exploration activities on fossil fuel
- the development of alternatives for fossil fuel
- the increase of energy efficiency
- and the demand for energy

will connect to each other. An overview on the different directions for lowering the gap between demand and supply of fossil fuel can be found in Figure 5.3 (Robinson and Mayo, 2008).

A situation of a 'supply constrained world' could arise and the period between 2015 and 2030 could become difficult 'the oil supply constrained world will be with us for much longer than currently anticipated by policymakers and will turn the current turbulence into our everyday affairs' (Jesse and van der Linde, 2008).

Giddens foresees, in a working paper for *The Politics of Climate Change* (2008b), great problems:

---

16   For a short and concise analysis, see Lutsey and Sperling (2009).

A situation where the world energy supplies peak relatively soon could be disastrous for world stability and for hopes of containing climate change. Modern industrial civilization is very heavily based upon oil and gas, not just so far as energy (especially transport) is concerned, but because they figure in a massive diversity of manufactured goods upon which our everyday lives depend. Over 95 per cent of the goods in the shops involve the use of oil in one way or another. There is simply no way of breaking this dependence in the short run, however successful we may be in reducing it. *A serious and prolonged shortage of oil might lead to economic chaos not only in the countries or areas of the world directly affected by it, but in the world economy as a whole.* (Giddens, 2008, 15)

In 2008 Shell presented *Energy Scenarios 2050*. Shell calls this period until 2050 'a time of revolutionary transitions'. The developing countries will need much energy in their take-off phase of economic growth, the CO_2 problem must be tackled, and the supply of oil and gas will become poor '. By 2015, growth in the production of easily accessible oil and gas will not match the projected rate of demand growth' (Shell, 2008, 8).

Shell presents two energy scenarios: *Scramble* and *Blueprints*. *Scramble* is a scenario in which bilateral agreements between producers and consumers dominate. National government are in concurrence regarding the scarcer resources. The oil and gas producers set the rules. The world that arises leads to a situation where international relations are mainly a race to ensure continuing prosperity. Tensions between countries will arise and governments will look at already known alternatives for their energy demands. Coal will become popular again, and the same holds true for nuclear energy. The disadvantages of more CO_2 emissions and of sensitivity to terrorism are taken for granted. The process is described by Shell as 'first, nations deal with the signs of tightening supply by a flight into coal, heavier hydrocarbons and biofuels, then, when the growth in coal can no longer be maintained, an overall supply crisis occurs, and finally governments react with draconian measures – such as steep and sudden domestic price rises or *severe restrictions on personal mobility'* (Shell, 2008, 20). It is clear that in Scramble the focus on the battle against global warming is minimal. Much time will be lost.

Blueprints is a much calmer scenario where the global interrelations are acknowledged from the start. From cooperation between the urban regions all over the world a spectrum of initiatives is being developed. Out of these initiatives a broader cooperation arises, leading to global agreements. The interrelations between security of delivery of energy, certainty on demand, and CO_2 management are seen and taken care of. There comes a CO_2 price mechanism with tradable emission rights worldwide. In exchange for their contribution on CO_2 reduction developing countries receive technology transfers and a greater share in the scarce oil and gas supplies, that have been broadened by OPEC, now certain of the demands. No time is lost in Blueprints and the transition to more sustainable energy can take place.

Scramble relates to the analysis by Jesse and van der Linde. There is a strong resemblance to a scenario painted by (the daughter of) van der Linde. In 2007 she wrote an article *Energie: de eeuw van mijn moeder* (Energy: the century of my mother) in which she looks back from 2040. In 2027 a great energy crisis arose, mostly because the energy delivery problems of the world seemed unsolvable. Global warming has totally been forgotten as a wider problem. After the crisis of 2027 it became possible to work on an energy transition. To quote 'and so an end did come on the restrictions on mobility and on the productions of goods that were quite normal in the times of my mother, but had become luxurious in my time' (Linde, 2007).

At the moment Scramble looks more realistic than Blueprints. Blueprints asks for coordinated efforts, for subtle diplomacy, and for acknowledging mutual interests in a way that has not been found yet. In *Signals and Signposts* (2011, 69) Shell states; 'Many people hope developments will be "faster than Blueprints". But at this point, developments are generally proceeding "slower than Blueprints", despite some achievements. Looking ahead, economics volatility and cyclicality threaten to depress the pace of change still further.'

This means that we have to take into account the coming of a turbulent period, in which fossil fuel will not be available for all objectives in the OECD world. Restrictions on mobility and more in general, a far more conscious use of mobility will become necessary. From our actual day-to-day practice this looks rather revolutionary!

5.5 CO_2 and Car Mobility

Car use and broader, transport, contribute to CO_2 emissions, and hence to global warming. In the Stern Report is stated that transport is worldwide responsible for 14 per cent of the global CO_2 emissions. The majority of these emissions come from road transport (76 per cent) and air transport (12 per cent). Forty-five per cent of all transport emissions comes from passenger cars (normal cars plus small pick-ups).

Production of these emissions is dominated by the OECD countries. In 2008 they were responsible for 4.1 of the 6.5 gigatons of transport emissions. This is 63 per cent of the total emissions, with 37 per cent coming from Canada and the US, 19 per cent from Europe, and 7 per cent from Japan, South Korea, Australia and New Zealand. With unchanged policies a growth to worldwide 9.3 gigatons in 2030 is expected (42 per cent more), and a growth to 12.0 gigatons in 2050 (91 per cent more). Transport will then still be responsible for 14 per cent of the total of emissions.

In OECD countries the share of transport in the production of CO_2 emissions is far higher than these 14 per cent. In the US 27 per cent of emissions came from transport (AASHTO, 2008), in Great Britain 24 per cent and in the Netherlands 23 per cent. The share of transport in the total spectrum of CO_2 emissions in OECD countries is still growing.

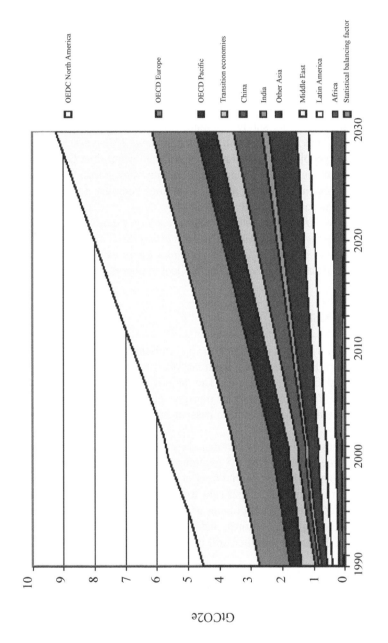

Figure 5.4   Transport emissions, parts of the world, 1990–2030
Source: IEA (2008a). Worldwide trends in energy and efficiency.

Concentrating more on Europe, as we saw with a share of 19 per cent of all transport emissions worldwide. This means that Europe's emissions are now 1235 megatons. The European Commission expects a rise in all transport sectors (minus shipping) from 970 megaton (2000) to 1093 megaton in 2030, meaning a growth of 14 per cent, rather low compared to the worldwide growth in transport emissions of 42 per cent in this period.

5.6 CO_2 Objectives on transport and mobility

The figures presented thus far were policy neutral. There is however a start with the policy on CO_2 emissions from the transport sector. In academic circles and among policy analysts the need to reduce CO_2 emissions in general by an order of 60–80 per cent in 2050, compared to 1990 levels, is anathema. The objective for 2020 is mostly 20 per cent, where possible through joint effort higher to 30 per cent, compared to the level of 1990. The official EU policy is; at least 20 per cent, and above that, the EU 'is willing to accept reduction to 30 per cent of greenhouse gas emissions in 2020, compared to 1990, under the condition that other industrialized nations are willing to accept the same order of emissions, and the economic better equipped developing countries will take a share in relation to their responsibilities and capacities'.

This objective has to be translated to all societal sectors. However, there is as yet no specific objective formulated for transport. Often it is acknowledged, as in the Stern Report, that 'transport will be among the last to bring its emissions down below current levels.' The reasons for this late reduction are 'the low carbon technologies tend to be expensive and the welfare costs of reducing demand are high'. However, Lutsey and Sperling (2009, 228) reject this view. In their opinion transport can make an earlier contribution. They estimate the advantages of lower energy costs (through higher fuel efficiency) as higher and the expected societal resistance as lower than the Stern Report.

The first reports on what needs to happen in Western European societies in order to reach reductions of 60–80 per cent have now been published. In 2008 the British Parliament passed the Climate Change Act. This Act set binding targets of a reduction of 26 per cent in 2020 related to 1990, and an 80 per cent emissions reduction in 2050, related to 1990. In the same year Buchan presented in *Low carbon transport policy* a first analysis of what these objectives would mean.

In the Netherlands in 2007 *Green4Sure* prepared a translation of a global objective of 50 per cent emissions reduction in 2050. Transport did get a share of 35 per cent reduction in 2030 (26 megatons), this time compared to the 2005 level (40 megatons). Even with this relatively limited objective, in this analysis it looks like more will be necessary than just technical measures. Transport can start reducing emissions slowly but has to end with determination. The real emission reductions from transport will in the view of Green4Sure come in the period 2030–2050. For the short term the reduction of 20 per cent looked reachable, however this is now brought into question. This 20 per cent general reduction means in the Netherlands that transport should stabilise in 2020 on the level of 1990 (32 megatons). This looks simple but this perspective changes when the Dutch figures are presented. Dutch transport growth between 1990 and 2008 was about 26 per cent. CO_2 emissions from transport have grown somewhat more over the same period, with 28 per cent (in 2008 nearly 38 megatons).

In almost all Western European countries these types of counting on objectives is being done in this decade and in each country the same three types of measures and instruments come into view:

- reduction of energy use per kilometre travelled, by realising energy/ fuel efficiency
- reduction in the (growth in) kilometres travelled
- adopting new vehicle technology, with lower greenhouse gas emissions.

From a distance a convergence in visions can be seen and there seems to be consensus on the direction of the policy. The differences arise with more pessimistic or optimistic estimates about the speed of the technological change needed, the progress in consciousness of car users and consumers, consumers, and expectations on the willingness and the power of European and National governments to reach agreement over high emissions reduction levels, which can be used as milestones, also when societal resistance to those reduction levels will grow.

5.7   Energy efficiency and vehicle technology

After 2020 in passenger transport a reduction of greenhouse gas is needed. At the same time, regarding car mobility, a drop in the dependence on fossil fuels seems necessary. For both objectives real substance is necessary as other sectors of society will probably not tolerate transport lagging behind too long. In principle two changes could create such substance; raising the fuel and energy efficiency and a serious shift towards less or no CO_2-emitting cars, using very little or no fossil fuels.

Energy efficiency

To reach objectives in the order of minus 60 per cent in 2050, compared to 1990 levels, it will be essential to use every opportunity for creating energy efficiency in passenger car transport. In the nineties there was a growth in energy efficiency but this growth was offset in most OECD countries by the trend towards buying and using heavier cars, SUVs and pick-up trucks. No leakage of energy efficiency towards higher car performance, by creating heavier cars, this should be the motto. This demands measures within the global car industry.

The European Commission negotiated covenants with the European, Japanese and Korean car industries whereby, starting in 2008/2009, newly-built passenger cars on the European market should have average emissions of 145 g/km, compared to 185 g/km in 1995. The European Commission also proposed to create legislation forcing car manufacturers to reduce emissions from cars to 130 g/km, working to create a emission level in new passenger cars in 2012 not higher than 120 g/km. There is strong opposition to these proposals, especially from the car industry, which dislikes the tone and the concrete actions.

In recent decades the energy efficiency of cars has grown substantially. Annema, Hoen and Geilenkirchen present figures in *Review beleidsdiscussie CO_2-emissiereductie bij personen vervoer over de weg* (2007). They note that in principle there is a great potential for energy efficiency. Twenty to 25 per cent less energy use from 2005 to 2020 should be possible. However, the real progress made in recent years was not impressive. Between 1998 and 2005 the emissions factors of recently-bought new cars diminished by 7 per cent, to an average of 170 g/km. The reason for this limited reduction, and this is the case worldwide, is that the greater part of the technical progress on energy efficiency and economy in cars is offset by a huge increase in the purchase of heavier cars, using more energy. The authors argue that creating chances for energy efficiency will, from now on, mean setting robust targets, and at the same time motivating car buyers to buy their cars in the non-heavy segments.

What will then be possible is presented in two reports. The first report, *On the Road in 2020: A Life-cycle Analysis of New Automobile Technologies* (Weiss and Heywood, 2000) sets the tone. The conclusion from this report is that continuous development of traditional petrol-based car technology could create cars a third more efficient in energy in 2020, creating a third less CO_2, and being only 5 per cent more expensive. A lighter petrol engine with lower resistance will then be created. According to the authors 'to move to most of these new technologies will require a change in customer behaviour, whether forced by government or voluntary '(Weiss and Heywood, 2000, 25).[17] The authors also put forward the idea that after 2020 some 50 per cent energy efficiency will be possible, with cars being 20 per cent higher in price.

The second report, *Factor Two: Halving the Fuel Consumption of New U.S. Automobiles by 2035* (2007) takes a further step. Halving would be possible by:

- substantial improvements in traditional car technique
- great reductions in car weight
- introduction of hybrids (fossil fuels and electricity).

These three elements are interdependent 'the weight, performance and fuel consumption of future vehicles are therefore dependent upon how improvements to conventional automotive technology are utilized' (Cheah et al., 2007, 14). The authors show how a reduction of 50 per cent on efficiency would be possible in

---

17   Note that, in 2012, we not on this track in Western Europe. The developments are proceeding slower than Weiss and Heywood advised.

2035, but this asks for the acceptance that in the in-between period of production higher CO_2 emissions will be at stake. And the start should be immediate; it would require that two – thirds of the new vehicles production be hybrids, requires 75 per cent of the energy improvements to go into fuel reduction instead of into increased performance and size, and requires a 20 per cent weight reduction (Heywood, 2008, 11). Cars will be 20 per cent more expensive and somewhat less comfortable.

Vehicle technology

Changing vehicle technologies is the longer term issue. Work is being done on the reduction of the dependence on fossil fuels, from the strategic notion that being dependent on a single commodity creates vulnerabilities. For transport especially the second generation biofuels, electric powertrains and hydrogen will be relevant.

Kromer and Heywood present in *Electric Powertrains: Opportunities and Challenges in the U.S. Light Duty Fleet* (2007) an expected scenario. Until 2030 the conventional powertrains will dominate, with ever greater efficiency. Hybrids get higher market shares, and a substantial part of car markets from 2030 onwards, on the premise that government actions will be there to reduce the reluctance of car consumers to pay more for their cars. Hybrids are an intermediate technology. Full electrical vehicles will really penetrate the car market after 2030.

Mourik carried out a study on the expectations and judgments of stakeholders –scientists, environmental specialists, energy experts and government officials – on transport energy policies. In their *E-managing Future Energy Infrastructures in the Dutch Transport Sector* (2005) it was concluded that scenarios with greater behavioural changes will be far more difficult to implement than scenarios based on technical renewal. Stakeholders expected diversification, see that the first phase in energy reduction will be based on improvements in efficiency, but differ on the role of biofuels, hydrogen or electricity. Creating a sense of urgency will be necessary in all relevant organisations and institutions. However, this being stated, they see great difficulties in creating this sense of urgency when there is not even a defined winner of the game of shifts in car and powertrain technology.

In recent times Western Europe has experienced a great deal of hype about the electric car. This is interesting. Experts all see that it will take at least 20 years before the domination by cars using fossil fuels ends and probably longer (see for example Heywood and Bandivakhar, 2005). However, the idea has been created that electric cars will be with us in vast numbers rather soon. Take the Netherlands as an example. The line of argument is as follows: hybrids now (2011) have a market share of 3 per cent, and this market share will grow in 2020 to 20 per cent, meaning that, yearly, 1 out of 5 car buyers (on average 550,000 cars bought in the Netherlands annually) will buy a hybrid.[18] The fully electric car will be

---

18   This sounds reasonable but it is also important to take into account that more than 50 per cent of Dutch car buyers buy second-hand cars. These second-hand cars are seldom hybrids. Nearly half of new cars bought in the Netherlands should be hybrids by 2020!

encouraged over this decade and will grow substantially in market share. In the *Plan on Electric Driving* of the Dutch Government (*Plan van Aanpak Elektrisch Rijden, Ministerie van Verkeer en Waterstaat, bijlage* 5, 2009) it is stated that fully electric cars may have a market share of 13 per cent in 2020. This will then mean that around 2020 a third of all cars will be hybrids or electric cars. Then, after 2020, changes will occur more quickly; the battery technique is then developed. On this theme the *Dutch Planbureau voor de Leefomgeving* has to say 'it can be expected that between 2020 and 2040 the cost of electric driving will be lower than the cost of conventional of hybrid cars. The price of the batteries is responsible for the greater part responsible for the price of the electric car.' (PBL, 2009c, 10) The Dutch government sees take-off for the electric car between 2020 and 2030; the market share goes up from 13 per cent to 69 per cent!

Looking at the huge number of problems that needs to be solved[19] before we have a real business case for electric car mobility these figures look somewhat strange in an official government paper. The explanation is probably the wish to present an alternative to a stagnating system,[20] of which can be expected that it will not be there in the near future.[21] And yes, electric car mobility can create new horizons to travel to. This '*deus ex machina* aspect' of the electric car is not immediately problematic but can become a burden if decision makers stop thinking about other alternatives on car mobility because they believe they have found the winner.[22] At the moment the full electric car is not more, and not less,

19   To name a few; financing the infrastructure for battery recharging, the changes in and on the road infrastructure, the number of kilometres that can be driven before charging batteries, the scale of extra electricity use, the CO_2-friendliness of the produced electricity for the cars, the protocols for peak use. Many experts see full electric cars as cars for the urban environments, as the second car, and not as a real alternative for the conventional car (see for example, Vleugel, 2009).

20   Aigle et al. (2007) on this issue; the technical core of the fossil fuel run motor has remained unchanged for a long period. Forms of hyperselection will than arise; a stable situation is created, the dominant technique is too strong and has created so many interdependencies with suppliers, academia and enterprises, that this technique becomes more or less immune to bigger changes.

21   Canzler and Marz called (1996, 17) the conventional motor and the related R&D investments an example of 'stagnation'; The stagnation of the Automobile Pact is innovative, in the sense that all intellect is used to search for alternatives within a very limited search space ('*die Stagnation des Automobilpaktes ist innovative, insofern alle verfugbare Wissen mobilisiert wird un die vorhandenen aber eng bemessenen Spielraume bis zum aussersten auszureisen*'.)

22   Luijendijk, a well-known Dutch journalist, made a series of articles on the shift to electric cars. One important statement; 'the reality is that all experts that I spoke have said: this is a transition of 30 years, and there are too many uncertainties to be able to say which technology will win (biofuels, electricity, hydrogen, very economic conventional cars). It will be a mix, so let us stop with ideas like: 'The Netherlands goes full speed for electric mobility!' (NRC, 9-1-2010)

than an interesting option, which could be developed out of the hybrid motor. The hybrid should then have a substantial market share in 2020.

In 2007 Aigle and Marz published a German study on vehicle technology with the title *Automobilitat und Innovation*. With an innovation matrix they show that at first incremental improvement will be at stake, with some enthusiastic activities on alternative power trains. There is a variety and there are no defined winners. However, some 'bridge-technologies' can be defined, technologies creating bridges between that of the conventional motor and the new technical core of the car of the future. The hydrogen motor and the hybrid are seen as these bridge-technologies (Aigle and Marz, 2007; Aigle, 2007)

5.8   A perspective on change; the role of car drivers, enterprises and governments

As we have already touched upon there are a number of reasons for the slow introduction of important changes in the transport sector. At first, car drivers have chosen heavier cars, thus lessening the effects of higher *energy efficiency*. They have on average chosen petrol cars, rather than diesels which emit less CO_2. They have chosen safety (airbags, and the idea that 'heavier is safer'), comfort (air conditioning, gadgets) and speed because they could. The car industry made their cars more energy efficient but kept offering the complete spectrum of cars. As consumers have chosen the heavier models, there was a leakage effect on the technical improvements. Between 1998 and 2005 the capacity of new cars grew by 19 per cent and their weight by 11 per cent (Annema et al., 2007, 19). The net effect of these choices has been that in the US improvements in fuel efficiency were lost through this growth in capacity and weight. In Western Europe more than half of this improvement has been lost, as European car drivers choose SUVs and small trucks far less often than their American contemporaries.

Secondly, on car technology we see a situation which will make introduction of new technologies a longer term issue:

- the techniques are not completely developed
- there is doubt about the winner, and hence doubt about where to invest
- the infrastructural changes to facilitate cars with other power trains are expensive and differ according to the chosen techniques
- the effect is only substantial when the new power trains will be used throughout the complete car fleet. Car fleets in OECD countries are changed completely in a time span of 14 to 17 years. A third of these car fleets are changed in 5 years (mostly lease cars and business cars), some 40 per cent is changed between 5 and 10 years, and the last quarter is changed after 10 years. This means that after a complete business case (which is not yet shown!) it will take some 15 years before an introduction is implemented

for 90 per cent. Before 2030 we should not have expectations on complete, or nearly complete, coverage of new vehicle technologies

- the introduction of radical, or paradigmatic other power technology is rather difficult because conventional technology changes in the change time needed also in the proper direction. The advantages of the really new technology 'damp away' before these technologies become market proof!

These reasons combined are responsible for the strategy of most car companies; certainly investing in R&D on newer, more radical, technologies but not on a really comprehensive scale, although their external communication often sounds otherwise.

Looking at the targets for CO_2 in the longer run (60/80 per cent reduction in 2050) and looking at the wish to become less dependent on fossil fuels, the task from now on will be a heavy one. Car consumers have to buy perpetually the cars that emit least CO_2, producers need to make these cars very soon and in huge numbers. Only when consumers and producers join forces in buying and creating the best possible cars, with the best available, CO_2-friendly, non-fossil fuel technologies, will it be possible to reach the CO_2 targets and to achieve less dependence on fossil fuels. To cite Chea et al. (2007, 34) on this theme 'it will require a new set of policies that pushes industry to utilize new technologies, while at the same time creating market demand to pull efficiency gains toward reducing fuel consumption and aligning the interests of diverse stakeholders to realize this worthy and ambitious goal.'

Is success possible? I will make a short analysis of the key players. First the car drivers, the *consumers*. Energy efficiency of car mobility is not very high on the consumer's priority list.[23] Higher fuels prices create, as we saw in 5.3, some extra sensitivity of consumers to the reduction of their energy use. Until 2030 raising the level of fuel efficiency will be the basic strategy for CO_2 reduction in transport. This is only possible when consumers really shift to the most energy efficient cars. The current situation here is presented by a manager of Ford Europe 'customers won't pay a premium for fuel economy. It is a mid-level concern for customers, not in the top three, but in Europe not nine or ten as in the United States.'[24]

In Western Europe the idea is that CO_2 levels should be reduced and the relationship with the economics in car driving is acknowledged, but no urgency is felt. In the US the situation seems slightly different. A study called *Consumer Views on Transportation and Energy* (2003) by the National Renewable Energy Laboratory states that Americans feel urgency related to creating lower dependence

---

23   As Whitmarsh and Kohler (2010) identify, consumers are aware of the problems, but this awareness is not manifested in low carbon travel behaviour.

24   Along the same line Kurani and Turrentine (2004); consumers miss the basic information to make real informed choices on fuel efficiency; 'when offered a choice to pay more for better fuel economy, most households were unable to estimate potential saving, particular over periods longer than a month.'

on oil from the OPEC countries, however, they feel no urgency at all on CO_2, and no urgency is seen in their car buying behaviour; only 30 per cent prefers lighter, energy efficient cars over average cars.[25] The equivalent percentage for the EU countries was, in 2003, around 62 per cent.

In *Clean and Efficient Vehicles: A Review of Social Marketing and Social Science Appliance* (2003) Kurani and Turrantine offer an explanation for the behaviour of consumers. They note that policy makers approach car customers as rational decision-makers, who will, all arguments heard, do the best for society. In their view these decision-makers more-or-less deny the complex and intricate decision-making process of car buyers, on the one hand not denying global warming, but being sceptical about the time scales presented, thinking it can and should be mitigated, but not wishing to lose their customary comfort and safety levels too soon.

The authors consider it necessary to restructure the car market; it does not help consumers to buy energy-friendly cars, while there remains the possibility to buy rather CO_2-unfriendly cars. In such a market 'free riders' are created, effects of good behaviour are being lost, and mistrust among car consumers arises.

Another element is that 'cost savings do not motivate consumers'. From their research it become clear that people will buy energy friendly cars as their part in the battle against global warming, and more widely, for non-specified environmental reasons, but not to save some dollars or euros. The last important element is that most consumers do not make big 'consumption jumps'. Their behavioural choices tend mostly to be step by step. More social marketing seems to be needed for energy efficient cars. Pessimism is not necessary but realism in time schedules on car buying is!

The *producers* look at their sales and notice the hesitation of the consumers. They react along two different lines; business as usual, presenting models with extra energy efficiency and sustainability elements, and, the other line, in their publicity an emphasis on R&D, on preparing for a shift to a new generation of cars, and on a few models that are really different, but do not make the bulk of their production lines.[26]

Producers want to have a good press in their contribution to the fight against global warming, but at the same time need the revenues of 'business as usual' (with a little more sustainability).[27] For most car industries it is significant that

---

25   This seems to change somewhat in recent years.

26   In this respect Wells (2006, 292) asks the question whether the car industry will be able, after a century of 'fossil fuel business models', to make a fast shift towards new models; 'has the prevailing business model has its day?'

27   Annema et al. (2007, 24) analyse; the car industry has invested so much now in clean cars that it cannot be called 'window dressing' anymore. At the same time the bulk of these investments of the car industry are defensive and risk averse. Reason: 'dinosaurs don't fly'. And Nieuwenhuis (no date, 1) calls the attempts of the car industry towards sustainability; 'rearranging the ashtrays on the Titanic'.

'given the high level of uncertainty concerning climate science, technological and market developments, and policy responses, car makers cannot easily make a rational, objective calculus of their economic interests and appropriate strategic responses, and might therefore be more subject to institutional pressures'(Levy and Rothenberg, *Heterogeneity and change in environmental strategy; technological and political responses to climate change in the global automobile industry,* 2002, 1175).[28] Within the car companies different views can be seen; mostly the R&D departments are in favour of acknowledging and anticipating global warming, while sales departments are rather reluctant!

All car companies see the consumer as their biggest problem. They wonder whether the investments needed for innovation to relate to future CO_2 targets will be cost effective, looking at the rather reluctant behaviour of their customers. A Volkswagen manager, quoted in Levy and Rothenberg said 'we are more active in trying to change consumers. You cannot force people to buy certain things, but what we are trying to do is just to keep on presenting it to the market, and try to convince people to buy these type of products.'

But car producers are also hesitant themselves. Orsato and Wells show in *U-turn: The Rise and Demise of the Automobile Industry* (2007, 1005) that car producers know their lock-in; 'automakers have responded to the regulatory and market pressure but the technological paradigm orientating car design and manufacture substantially limits the alternatives available to them – perhaps explaining the more positive attitude towards hybrid powertrains compared with fuel cells'.[29] Having so many 'sunk investments' creates a bias among producers towards slower, incremental change.

Policy makers should look at strategies that are related to the institutional contexts of car companies. Trusting market-related instruments is not enough, market instruments have to be framed in inspirational visions and directions about market developments and necessities to regulate car markets.

This brings us to the third party involved, the *policy makers.* Consumers see the need to change, but hesitate. Producers see a scope for change, but hesitate. Are policy makers able to break the impasse that will arise? There is probably a need to convince car drivers to buy energy efficient cars, and there is a need to change the type of cars that can be bought in a sustainable way. This asks for genuine market regulation and there is need for permanence in vision and measures,[30] like the norms for CO_2 efficiency of cars. By these elements

---

28   There are rather great differences between the car companies. Until recently, American car companies tended to deny global warming, whereas most European companies accepted global warming and related their policies with the European government policies (Levy and Rothenberg, 2002, 1170).

29   A similar argument is elaborated in *Structure and Agency in the Car Regime* by Marletto (2010).

30   Here is a problem of attitude, as Sandalow (2008, 3) states; 'Policymakers eager to see dramatic changes in oil consumption will find the pace of change remarkably slow.'

combined the producers feel a climate coming in which they dare to invest on a really substantial scale in the production of the innovative and energy-friendly segment of the car market.

This all asks for endurance and for policies related to diminishing the amount of car kilometres driven, that will be a lasting element in policies to fight global warming. As Lutsey and Sperling (2009, 227) conclude 'achievement of 2050 climate change stabilization will almost surely require substantial reductions in per capita travel and a new round of more efficient and low-carbon vehicles and fuels'. At this moment policy makers are not ready for these measures. We still are in an early phase with a hope on reaching 60–80 per cent targets with only technical measures.

Now the objectives are described. Lobbies are influencing time schedules to reach these objectives, and systems for CO_2 emissions trading are being set up. Finally companies will have to pay for the right to create emissions and these companies will make their consumers partly pay for their higher costs. This all demands political leadership.[31]

CO_2 reduction of passenger car mobility will be found in the complete functioning of the triangle consisting of car consumers, car producers and policy makers. What will then be possible is shown in an overview prepared by Heywood and Bandivakhar. In 2005 they presented an article on *New Vehicle Technologies: How soon can they make a difference?* In their vision the process of market penetration has three distinct phases. The first phase consists of the development of the technique until the moment of presenting to the market. The second phase is the market penetration *sensu stricto* and the third phase starts at the moment when the new technology has become the core technology to be used.

The authors conclude 'the point is that these time scales are all long, and some are very long. It adds urgency to the fact that we should start trying to prompt these charges right away' (Heywood and Bandivakhar, 2005, 4).

5.9 Scenarios for car use, sustainability and energy

In this section we will look at the two problems jointly. We noticed that in the period 2010–2030 there will be the possibility of problems delivering fossil fuels, and there is a need to stabilise CO_2 emissions from passenger transport by 2020, and later to strive towards huge decreases. In other sectors of society CO_2 emissions reduction is already the current situation, but in transport we see

---

31 This more being the case because Petersen and Andersen (2009, 17) show that radical changes, not directly related to day to day practices in normal life, will demand greater governance capacities; 'changing practices in relation to everyday (car) mobility in favour of environmental considerations will almost inevitably interfere with other everyday life considerations, and the larger the interference, the larger the potential barriers to the shift.'

still some emissions growth. Other societal sectors are, at the moment, helping transport. The big changes for transport have been identified, there are initiatives but at the same time leakage effects and hesitation to buy or to invest can be seen. Energy efficiency will be realised but at higher costs for car users, and using less comfortable, lighter cars. Substantial changes in vehicle technology will take place after 2030. Generally stated; the future on transport does not look bright, yet.

To show the effects of the questions regarding fossil fuels, CO_2 and sustainability in relation to car use I will present eight possible scenarios. For these I will work with four themes:

Table 5.1   Scenarios on oil and climate in relation to car mobility

Peak oil	Energy-efficiency Conventional cars 2010–2030	Electricity Take-off 2020 or 2030	Reaching strong CO_2 targets	Need for a non-technical package	Name of the scenario
Around 2020	huge	imminent	yes	no	Swift Shift
	huge	not yet	yes	yes, temporarily Related to delivery problems	Energy-efficiency as the basis, with a slight problem
	rather small	is coming	no	yes, to reach the CO_2 target	Narrow Escape, but later problems
	rather small	not yet	no	yes	Early Crisis
Around 2030	huge	take off successful	yes	no	Everything works
	huge	not yet	yes	yes, for the delivery problems	Energy-efficiency as the basis, only minor alternatives
	rather small	take off successful	no	yes, to reach the CO_2 target	Electricity is the basis
	rather small	not yet	no	yes	Late and Severe Crisis

a.   Peak oil; will peak oil be reached at an earlier date (around 2020) or at a later date (around 2030)? When peak oil arises problems of delivering fossil fuels can be expected; the amount of oil to be produced remains stable, the demand still grows.

b. Energy efficiency; will a huge package on energy efficiency of the conventional motor be reached in the next two decades (33 per cent plus), or will the results be more in line with the last two decades (20 per cent minus)?
c. Vehicle technology; will there be an earlier take off on electricity (hybrids and full electric around 30 per cent in 2020), or will the take-off take place after 2030?
d. Reaching huge CO_2 targets; will the transport sector reach the expected target for 2020 (stabilisation) and 2030 (decrease on a relative low CO_2 emission level) with the spectrum of measures as foreseen, or is this not the case?

From the scores on a. to d. the urgency of a package of non-technical measures can be identified. This package has not yet been introduced in this chapter. Basically the measures are aimed at reducing (growth in) car use. It is still rather quiet on this area, but the measures one can think of have remained unchanged for already 20 years more and a lot is written about these measures, especially in academia and among environmental lobby organisations. They describe investing in a very good working public transport, encouraging more people per car per journey, about awareness in the choice for car use and they pay attention to teleworking, and to physical planning in urban and rural regions that supports wise use of car mobility. In Europe the spectrum of measures is widely known, and with different successes introduced especially in urban regions. In Canada, the US, Australia and New Zealand interest in this spectrum of measures is growing (see for example Salon and Sperling, 2008).

In the Dutch study *Green4Sure* some signals are given on this relatively quiet field, at least quiet seen from the perspective of forthcoming oil delivery problems and global warming.[32] A policy that really intervenes in physical planning and in modal choices seems necessary and this is more than the voluntary situation of the moment. Interventions on the volume of car mobility will be necessary, however the authors write 'The past has shown that it will not be easy to use these policies effectively for CO_2 reduction' (Green4Sure, 2007, 104).[33] The quiet situation is explained with 'because it is not clear how much has to be steered in this area. That is strongly dependent on the results of the technical possibilities on energy efficiency, biofuels and of the costs to mitigate in this sector in relation to other societal sectors' (Green4Sure, 2007, 104).

---

32   Many interesting investments are taking place on the mentioned measures, mostly from liveability perspectives, and mostly in or around cities. However, the relation with the problems central in this chapter is seldom made.

33   For example in the American study Growing Cooler (2007) by Ewing et al. it is concluded that a 100 per cent increase in building density (twice as compact!) will result in 12 per cent fewer kilometres travelled in 2050, comparable with 7 to 10 per cent emission reduction. Creating a 100 per cent increase is a real paradigmatic change!

The eight scenarios together form the space in which probably over the next two decades the development of car mobility in a technical sense will take place.[34]

It is clear that when peak oil starts earlier the problems will be greater than if peak oil is reached at a later date. However, the CO_2 targets are on the earlier dates are not very strong. When peak oil starts around 2030 the situation could be; problems with the delivery of fossil fuels, but already a real market share of electricity, creating a smaller delivery problem, but a still great CO_2 problem.

What is called the non-technical package (or the societal package) looks necessary in most scenarios but for different reasons. The societal package is essential when energy efficiency and vehicle changes are not implemented early. The societal package is certainly necessary in situations with early peak oil. And there is a need for the societal package when after 2020 strong CO_2 objectives for transport should be realised, because other societal sectors urge for equity in approaches and targets. *To conclude, there is a rather great likelihood that between 2015–2030 realising greater independence of fossil fuels and realising realistic objectives on CO_2 will only be possible with interventions in travel behaviour and in the volume of car mobility. From this perspective it does not look wise to accept a further growth in car dependence.*

5.10   Scenario-building exercises in five countries

The objective of CO_2 reduction in a range from 60–80 per cent reduction (from the levels of 1995/2000) creates enthusiasm among transport researchers in different countries. Scenarios for reaching this challenging objective for transport are being designed. In this section I will give a short introduction to the spectrum of considerations and views.

I will start outside Europe. In a keynote lecture given November 2010, Sperling (Institute of Transport Studies of the University of California), presented his *Steps into Postfossil Mobility.* His comprehensive plan consists of measures in three so-called arenas; vehicles must become far more energy efficient, the carbon content of fuels must be greatly reduced, and consumers and travellers must behave in a more eco-friendly manner. He considers this last arena the most difficult. 'Cars are firmly entrenched in our culture and modern way of life. Reducing inefficient car-dependent vehicle travel requires reforming monopolistic transit agencies,

---

34   However, here one remark has to be made. One can think that when passenger mobility and especially car mobility will not succeed in reaching CO_2 reduction targets a discussion will start, with the outcome that there will be no choice for a package of strong and severe non-technical measures, but a choice to accept a lower target for car mobility. In such a situation there are two possibilities; other sectors of society have to reach higher targets, or the whole policy on climate change and global warming will be slowed down. In Chapter 7 this will be discussed in more detail.

anachronistic land use controls, distorted taxing policies, and the mindset of millions of drivers who have been conditioned to reflexively in the car every morning' (Sperling, 2010, 3). He designs and defines for each arena a specific set of measures but warns; 'Achieving a 50 to 80 per cent net reduction in greenhouse gas emissions is not something that businesses, consumers, and politicians can fully imagine. Life after cheap oil evokes images of crises to come. There is no escaping that there will be winners and losers, but strong leadership and good policy can ease the transition' (Sperling, 2010, 3).

In *Klimachutz im Verkehr-Paradigmenwechsel* (2009) by Becker, Clarus and Friedeman of the Technische Universitat Dresden, the magnitude of the task to reach the objective is shown in a creative way. Given the 60–80 per cent reduction objective, taking into account the carrying capacity of the world environment and equity worldwide in the share of fossil fuel to be used per person the authors show that each person (with 25 per cent transport use of all fossil fuel use) can be allowed to use 100 litres of fossil fuels in 2050. This means per person per year; 500–800 kilometres driving, 2000 kilometres flying, 10,000 kilometres on public transport, and ubiquitous walking or cycling.

In the UK, Buchan prepared *A low carbon policy for the UK* (2008). His conclusion is that 'policies which produce more efficient patterns of travel will be needed alongside those for improving fuel consumption both in the medium and the long term, and that they need to be implemented as a matter of urgency' (Buchan, 2008, 9). Buchan sees the need for land-use regulation, for behavioural change, with specific initiatives on shopping, the school run, work and leisure, walking and cycling, speed limits in car traffic, and on taxes.

The same line of thinking can be seen in Sweden. In *How much transport can the climate stand? – Sweden on a sustainable path towards 2050*, the authors look at the technical possibilities (Akerman and Hojer 2005). A far greater energy efficiency can be reached, however, the results of this success will disappear in the growth of car kilometres. The conclusion of their study is that 'a development towards sustainable transport requires significant changes in the organisation of daily activities and daily travel' (Akerman and Hojer, 2005, 1954). To reach the objectives total car travel volumes have to be cut by 32 per cent, more in urban areas and less in the rural areas.

For France, Lopez-Ruiz and Crozet have prepared in *Sustainable Transport in France; is a 75% reduction in CO_2 emissions possible?* (2010) three quantitative scenarios. It appears that in each of the three scenarios a 50 per cent reduction will be possible by 2050. However it will be more difficult to reach in *Pegasus*, a scenario promoting individual travel with strict technology standards. The other two scenarios create better results. In *Chronos* constraints on speed are introduced, and green multi-modality is promoted. In *Hestia* the relationship between physical planning and transport is elaborated; increase in densities is a key element, and the decoupling of transport activities and economic growth is promoted. Going further than the 50 per cent would require very big advances in zero emission vehicles. To cite Crozet '*au total, les grandes tendences dans les* *prochaines annees se resument ainsi; moins vite (en ville et sur la route), plus cher et plus concurrentiel*' (Crozet, 2010) (in total, the big trends in the coming years are: less speed (in the city and on the highway), more expensive, and more in concurrence with other modes).

To finish Sperling (2010, 14) has this to say: 'The days of conventional cars dominating personal mobility are numbered. There are not sufficient financial and natural resources, or climatic capacity, to follow patterns of the past.'

Chapter 6
Frequent Car Use: Societal Questions

6.1 Introduction

Modern Western risk societies are growing towards car dependence but what does this mean for these societies? Is it just a neutral development, or will it come at a cost? And then, costs for whom? Which persons, households, cultures will face extra problems as a result of further growth in car dependence and car-dependent arrangements in society?

These questions are central in this chapter. We will work in three widening circles, starting small and specific (with carless households) and ending broadly and in general (with the characteristics of car-dependent societies).

In Sections 6.2 and 6.3 the situation of households without a car will be analysed. The focus will be on facts, figures and the current situation of these households in Section 6.2 and on their mobility patters in Section 6.3.

In Section 6.4 the focus is broadened to all groups facing some form of transport or mobility disadvantage. Here we also look at poorer households forced to buy cars to be able to join the chances of modern societies. The relationship between transport, mobility and social exclusion will be explored.

In Section 6.5 the focus is broadened once more. Here we look at the relations between frequent car use, car dependence, and the social and cultural characteristics of modern societies at large. Which living patterns are stimulated by frequent car use or car dependence, and which living patterns are in decline or even disappearing?

The last part of this chapter will be used for two *capita selecta*. The first (Section 6.6) is on congestion, essentially a middle-class problem of too little car infrastructure at peak hours. Congestion is a function of the way we have organised our modern risk societies and a function of the results of our spatial and physical planning. And the second (Section 6.7) is on the shape of things to come. How will IT and the situation when the first IT-generation reaches adulthood changes perspectives on car mobility?

6.2 Who are the carless households in Western Europe?

This question will be answered in this Section along two lines. First, in some detail for the Dutch situation. Second, we will look at facts and figures from other countries. In the Netherlands in 2009 20.9 per cent of all households did not have a car.

Table 6.1   Households and car ownership, the Netherlands, 2009 (%)

	More than 2 cars	2 cars	1 car	No car
One person household	0.1	1.1	54.9	43.9
Two person household	1.2	23.9	64.4	10.5
Three person household	6.4	38.0	47.5	8.1
Four person household	7.2	46.9	42.5	3.4
Five person household	8.0	44.0	44.2	3.8
Six person household	10.1	45.4	42.4	2.1
Total of all households	2.6	21.6	54.9	20.9

Source: Mobiliteitsonderzoek Nederland 2009.

Of all single households, nearly 44 per cent does not have a car. Note that families with four or more members often have two cars instead of one.

Non-car ownership is strongly concentrated in single households. Of these single households, in the Netherlands 34 per cent of all households, 1.1 million do not have cars. Of the two-person households (29 per cent of all Dutch households) 10.5 per cent do not own cars. This means 230,000 two-person households do not own cars. And of the families (37 per cent of all Dutch households) some 5 per cent do not own cars, meaning 140,000 families. In total, a little below 1.5 million Dutch households do not own cars.

This means that some 14 per cent of the total Dutch population does not have a car to use: this is about 2.2 million people.[1]

More depth can be brought to these facts. Nearly 80 per cent of all men between the ages of 30 and 75 can use a car and 60 per cent of all women between the ages of 30 and 60. Most men without a car are young (18–27 years), women take longer to get a car (until on average 32 years of age) and the majority of women do not own a car once past 60. Looking at couples, the range is from just one car between 25 and 30 years of age, to nearly one and a half cars between 40 and 50 (Jeekel, 2011, 198).

Singles

Seventy-two per cent of all Dutch carless households are single households. Seventy per cent of these single carless households consist of one woman. Near to 50 per cent of these non-car owning single women are older than 60. Some 400,000 households without a car are older women, who mostly have never learned to drive. Some 20 to 25 per cent of the single carless households, 300,000, are young households (18 to 30 years).

---

1   More data in Jeekel (2011), although the author has used the 2007 statistics!

Couples

The carless households are here to be found in the youngest couples, 18 to 25 years, 80,000 households and in the eldest couples, above 75 years, some 40,000 households. The other half of the carless households is to be found in couples between 25 and 75 years old, mostly in cities.

Families

Very few families are carless. Children and cars seem to be interrelated. Single parent families have higher rates on non-car ownership. Whereas the national average is 76 per cent car share in kilometres travelled, this figure is 70.5 per cent in single parent households. The difference is even greater in comparing self-driving or being driven. The national average is 52 per cent and 24 per cent respectively, while with single parent families the figures are 38 per cent and 32.5 per cent.

Carless households in the Netherlands are mostly low income households (see Table 6.4, MON, 2009) and non-car ownership seems to be an urban phenomenon; in rural areas 11 per cent of the households is carless, in the four bigger Dutch cities 39 per cent.

The statistical profile of non-car households in the Netherlands is clear: Many singles, especially urban singles, young or elderly, including a substantial group of older women, young couples and the oldest couples, and single parent families. Looking at social groups the focus is on the elderly, some of the women, the young people, minorities and the disabled.

In Section 6.3 we will look at the car mobility patterns of minorities, disabled and the elderly. In most Western European countries there are now more households with two or more cars than households without a car. From Table 6.2 it can be concluded that 18–25 per cent of all households have no car. Forty-three to 55 per cent of all households have one car and 20–32 per cent of all households own two cars. More than two cars can be found in 2.6 per cent to 6.0 per cent of all households.[2]

---

2   Data sources are:

Sweden: RES 2005–2006 The National Travel Survey, SIKA (Swedish Institute for Transport and Communications Analysis), 2006.

Germany: Mobilitat in Deutschland 2008: Ergebnisbericht; Struktur – Aufkommen – Emissionen – Trends, Infas und DLR, Berlin, 2010.

Switzerland: Mobilitat in der Schweiz; Ergebnisse des Mikrozensus 2005 zum Verkehrsverhalten, Bundesamt fur Statistik/Bundesamt fur Raumentwicklung, Neuchatel, 2007.

France: La mobilité des Francais. Panorama issu sur l'enquete nationale transports et dep[lacements 2008, Commissariat General du Developpement Durable, 2010.

Flanders: Onderzoek Verplaatsingsgedrag Vlaanderen 3 (2007–2008), Universiteit van Hasselt, Instituut voor Mobiliteit, D. Janssen et al., 2008.

The rule of thumb for the selected seven countries is 21 per cent carless households, 49 per cent one-car households, and 30 per cent households with two or more cars.

Table 6.2   Ownership of cars, by households (%)[3]

Country	No car	1 car	2 cars	More than 2 cars
Germany	18.0	53.0	23.0	4.0
United Kingdom	25.0	43.0	26.0	6.0
France	19.0	45.0	32.0	4.0
Switzerland	18.8	50.6	25.1	5.4
Flanders	18.2	53.6	24.7	3.5
Sweden	25.0	52.0	20.0	3.0
Netherlands	20.9	54.9	21.6	2.6

Source: Different National Travel Surveys, see note 3.

Car ownership is still growing quickly in Western Europe. In the Netherlands car ownership in 1995 was nearly 5.7 million cars, compared to 7.6 million cars in 2009, a growth of 33.3 per cent or around 2.4 per cent yearly. Car density in the Netherland went from 390 cars per 1,000 inhabitants in 1995 to 460 cars per 1000 inhabitants in 2009. In the United Kingdom between 1994 and 2008 the number of cars went from 23 million cars to 31 million cars, a growth of nearly 35 per cent, or around 2.3 per cent yearly.

In France from 1994 to 2008 the numbers of cars grew from 26.4 million to 32.7 million, a growth of 24 per cent, and around 1.7 per cent yearly.

But which households have two cars, one car, or no car? In order to answer this we must take a quick look at the statistics in a few selected countries.

In *Germany* for example over half of the households with a net monthly income lower than 1,500 euros do not own cars, and households with net incomes higher than 3,000 euros have, in the majority, more than one car. Above 6,000 euros net income 25 per cent of households have three or more cars (Abbildung 3.34, 58).

When being asked for reasons not having a car 50 per cent of the non-car owners considered a car too expensive, 19 per cent gave health- or age-related reasons, 16

Netherlands: Mobiliteitsonderzoek Nederland 2009, Tabellenboek, Rijkswaterstaat, 2010

United Kingdom: Transport Trends, 2009 edition and National Travel Survey: 2009, Department of Transport, 2010.

3   In each table I use the figures of the country from the year of their National Travel Survey (2005, 2007, 2008 or 2009).

per cent did not need a car, and 5 per cent rejected cars (Abbildung 3.35, 59). Carless households are more to be found in the cities Berlin (41 per cent, Bremen and Hamburg, more in former Eastern Germany (around 25 per cent) and among singles (below 30 years 38 per cent, and above 60 years 46 per cent). Twenty-five per cent of single parent families do not own a car (Abbildung 3.38, 62).

In the UK the level of 25 per cent carless households has been stable since 2004. More than 43 per cent of the households in the lowest two income quintiles had no car. The majority of households in the highest two quintiles had two cars or more (62). Of the non-car owners 33 per cent said they were not interested in driving and 32 per cent could fall back on family or acquaintances. Cost factors have become more important in recent years, health and age reasons were important for 19 per cent. Carless households are to be found in the London boroughs (38 per cent) and in other major urban areas (31 per cent on average). Convenience was by far the most important reason for having more than one car and in rural areas more than half of all households had two or more cars.

In France, income is also important, but the location of the households matters even more. The highest car ownership rates are to be found in the rural and peri-urban areas, with very low rates in the bigger cities and extremely low rates in Paris (graphique 2, 104). The number of students, inactive people, unemployed and older people without a car is above average and 25 per cent of single parent families do not own a car. Carless households live on average far nearer to railway stations (104). The French situation is explained in *La mobilite, nouvelle question social* (Orfeuil, 2010). Thirty-nine per cent of the households earning less than 60 per cent of the average income do not have a car (Orfeuil, 2010, 6). In the rural areas a mere 12 per cent of all households have no car. For 22 per cent of all French households public transport is out of reach.[4] This means a rather high number of poorer households that need to maintain a car to reach locations of use to them.

An important article on this theme is *The Dynamics of Car Ownership in EU Countries* (2005) by Dargay and Hivert. They describe car ownership on an EU level. In the EU carless households are also often single households (45–81 per cent of EU carless households are single households compared to the EU single household average of 31 per cent), are more often older households (29–72 per cent of EU carless households are above 65 years of age compared a general average of 23 per cent), often do not have children (8–21 per cent of carless households compared to the general EU average of 30 per cent), are more often managed by a woman (29–57 per cent of carless households are managed by women compared to the general EU average of 14 per cent), and are often not working (19–81 per cent are carless households have nobody working in the household compared to the general EU average of 30 per cent).

The final part of this Section looks at the situation in the US. In 2001 nearly 8 per cent of all households did not have a car, with 5 per cent in the rural areas and 10 per cent in the urban regions. Of households earning less than US$25,000

---

4   Figures for Great Britain and Scandinavia: 15 per cent and 13 per cent.

a year more than 21 per cent did not have a car and of the 12 per cent Black households almost a third had no car (National Travel Survey).[5]

6.3   Mobility patterns of non-car households

The central question in this Section is how carless households take care of their mobility needs. A secondary analysis of the 2007 mobility statistics of non-car households in the Netherlands was made (see Jeekel, 2011, 203). These households travel fewer kilometres, only 60.5 per cent of the average number of kilometres in Dutch society. They make, however, more trips per day (3.21 journeys compared to 2.99 on average). A look at the reasons for this shows that the travel for work motive is in general far less important to Dutch carless households.

Table 6.3   Carless households in the Netherlands: Motives in kilometres and trips, 2007

Motive	Carless households	Dutch average	Carless households	Dutch average
	Distance %	Distance %	Trips %	Trips %
To and from work	22.2	26.9	18.0	17.3
Business	4.2	8.5	1.6	3.0
Personal care and services	2.6	2.8	4.9	3.6
Shopping	9.6	9.5	24.2	20.4
Education	7.3	6.1	8.2	9.3
Visiting	28.9	20.8	17.1	14.3
Social and leisure	14.2	12.7	11.8	12.0
Walking and recreational trips	9.1	8.7	9.9	10.7
	100	100	100	100

Source: Editing data Mobiliteitsonderzoek Nederland 2007.

Remember that the distance is already 60 per cent lower than average, of this lower distance the work motives explain 26.5 per cent compared to 35 per cent on average in Dutch society. On the other hand, carless households travel for visiting

---

5   These figures resemble the situation in the prosperous OECD countries outside Europe. 10 to 15 per cent of the households there have no car, with a bias on poorer and elder households.

nearly the same distance as the Dutch population at large and for shopping non-car households make more trips than average.

The modes that carless households use for their mobility are shown in Table 6.4.

Table 6.4   Transport modes used by non-car households, 2007

Transport mode	Carless households	Dutch average	Carless households	Dutch average
	Distance	Distance	Trips	Trips
Car as driver	4.1	49.4	1.4	33.1
Car as passenger	23.1*	26.3	6.5	15.4
Train	31.7	8.0	5.2	2.0
Bus, tram, metro	12.2	2.9	12.2	2.7
Bike	15.7	7.2	31.6	26.0
Walking	4.8	2.9	38.1	18.7

Note: * In the literature only little analysed; the idea that members of non-car households have to ask for being transported. In RAC Foundation (2009, 82); *The Car in British Society*; 'some people who did not own cars felt … that they were a burden on their friends and families'.
Source: Editing data Mobiliteitsonderzoek Nederland 2007.

In carless households the car takes 27 per cent of the share of kilometres travelled, mostly in one of the passenger seats, compared to 76 per cent on average. Public transport is far more important, as are the slow transport modes. It is interesting to note the 1.4 per cent of the trips and 4.1 per cent of the kilometres travelled by members of non-car households as car drivers. Twenty-one per cent of Dutch households have no car, but only 19 per cent have no driving license. In one out of ten Dutch households without a car at least one person can drive!

In general the question remains whether carless households face difficulties in their mobility. Can they make all trips that they want to make, with a special focus on those trips defined in Chapter 4 as car dependent? Acknowledging that working trips are mostly not so relevant for these households it is important to know how they organise visiting friends and relatives over somewhat greater distances, how they organise the weekly shopping trip, the journeys with a lot of luggage, and going on holidays. How do they reach the highway locations? There is a difference; the 10 per cent of carless households with a driving license can rent cars. For the other 90 per cent and for car-dependent journeys there are essentially three possibilities; to ask for a lift, to make a trip using a mode 'beyond what is seen as reasonable' (you can cycle 30 kilometres on Sunday morning for your sport competition!), or to forget about the journey.

For the non-car-dependent trips alternatives are available, in the slow modes, or in public transport. Not all carless households will have problems reaching destinations. We will look in more detail to the specific groups of carless households.

Starting with young households, which are mostly flexible about travel: They will probably buy a car eventually and few problems will arise. However, this picture needs some explanation. Choplin and Delage present, in *Mobilites et espaces de vie des etudiants de l'Est francilien; des proximites et depandances a negocier* (2011), an interesting case study on students in a difficult-to-reach smaller university. The university is located between Paris city, the suburbs and the rural areas. Students come from throughout the region, but most public transport does not offer services from their villages and home cities to the university directly. They can travel via Paris, but this is rather costly. They are very car dependent but only half of them have a driving license and only 28 per cent have a car. Each student searches for solutions to the problem but mostly the result is being driven by the parents after long negotiations, or using expensive public transport.

On the other side of the age spectrum, the elderly often face difficulties. Two groups can be identified. Older households that have never owned a car and households that have stopped driving. Scheiner shows in his study from Germany *Does the car make elderly people happy and mobile?* (2006) that car driving becomes more difficult for many people after they reach 75. Then 'it is the hale, healthy and therefore most satisfied and mobile seniors who frequently own a car' (Scheiner, 2006, 154). He shows that older households with cars engage in a greater variety of activities than non-car older households, which stay at home far more.

The Dutch researcher Tacken says the same in *Ouderen en hun Mobiliteit buitenshuis* (unknown year) and shows that cycling also declines after the age of 75. Older people walk more, and around 40 per cent of those over 75 years old do not leave their houses independently. Travel by public transport by older households is lower than expected. In his view decision makers overestimate the opportunities of older people to use regular public transport. The high costs, the difficult card systems, the problems with getting to and on the bus, the possibility that there will be no seats available and the fact that older people are sometimes anxious with encounters with strangers are all reasons to avoid public transport. Most older people prefer 'travel on demand', mostly serviced by taxis.

In a study from New Zealand, *Coping without a Car* (Davey, 2004), the way in which older people try to remain mobile was examined. Stopping driving was, especially for older men, particularly emotional. It is not an easy and straightforward process and older men have difficulties adjusting to life without a car. Older people who stop earlier with driving then have the necessity of developing arrangements for their mobility. Older men often lose the capacity to arrange their mobility in a satisfactory way.[6] Elderly households without car face the greatest difficulties in

---

6   This was also a theme in a Japanese study: 'It was also found that the elderly did undesired adjustment of their activity schedules to participate in medical care activity at hospitals' (Izumiyana et al., 2007, 15).

seeing family; it is that difficulty that makes them feel the most dependent (Davey, 2004, table 4).

Requesting and obtaining a lift becomes the most important form of mobility for the elderly, more important than the taxi and in New Zealand certainly more important than public transport. But older people are selective about asking for lifts. They rarely ask for lifts for leisure or social activities and as a result these activities can diminish. Very long friendships are no longer maintained because the friends cannot reach each other anymore. Lifts are requested for shopping and especially for health reasons such as seeing a doctor or travelling to hospitals. In New Zealand many older people do not leave their house anymore. Problems are not broadly mentioned; older people see this as a fact of modern life and they adjust accordingly.

In the Australian study *Maintaining Mobility* (2007, 20) it is mentioned that getting lifts will probably become more difficult 'Even where family members do live close by, they are often not as available as previous generations to assist with transport for various reasons. For example, a higher level of female participation in the workforce means less time for non-work activities. Many people are having children later in life and may have both young children and older relatives to look after ... the availability of private lifts may therefore be on the decline.'

Among the carless households many disabled people can be found. They form a rather specific category who, while being car dependent – via transport on demand, via taxi, via getting lifts – often do not own cars themselves. Regular public transport creates many accessibility problems for them as shown by Bakker and Van Hal in *Understanding Travel Behaviour of People with a Travel Impeding Handicap (2007)*. This is an inventory on the mobility of the disabled people, nearly 6 per cent of the Dutch population, nearly 1 million inhabitants. They make 30 per cent fewer journeys and travel 20 per cent fewer kilometres than average. Half of these disabled people are over 65. Younger disabled people need extra focus in mobility research. This risk group consists of 350,000 people, who are even less mobile than average; 25 per cent fewer journeys and over 50 per cent fewer kilometres travelled.

The figures for France are somewhat comparable. In 2010 Dejoux and Armoogum presented the current situation in *The gap in term of mobility for disabled travellers in France*. Ten per cent of the French population over 15 years of age (5.1 million people) reported difficulties when travelling outside their home. Their total immobility, measured by having gone out in the previous seven days, is far higher (19.2 per cent compared to the French average of 1.5 per cent). People mentioning difficulties make 20 per cent fewer journeys and these are far shorter than average. Fifty-nine point three per cent travelled for less than 30 minutes compared to 27.4 per cent of people without difficulties. The principle mode used among people reporting difficulties is the car, 57 per cent (lower than average, 69 per cent) but travelling as a passenger is higher than average 17 per cent (or 30 per cent of the total car use).

Another group with relative more carless households are the immigrant minorities. The Social Research Institute of the Netherlands SCP carried out a study on their mobility in 2005, *Anders Onderweg,* with some follow-up study (KiM, 2008). The most significant Dutch minorities – Turks, Moroccans, Surinamese and Antilleans – live mostly in the bigger cities of the Netherlands. Seventy-eight per cent of them live in the biggest 50 municipalities. The Netherlands has 1.2 million ethnic minority people (known as allochtones), 7.1 per cent of the population.

Allochtones make fewer journeys than the native Dutch population. For example, Turks and Moroccans make only 65 per cent of the journeys made by native Dutchmen in comparable situations. Turkish and Moroccan women in particular make very few journeys. The differences in distances travelled are even greater. Turks and Moroccans travel only half the distance, and stay in their cities. The Surinamese and Antilleans travel more, some 70 per cent of the distances of the comparable Dutch group. Mobility of allochtones is essentially urban. They do not use bikes very often, prefer walking and are relatively high public transport users. They have far fewer cars than average, with the exception of the Turks.

Single parent households as a group with lower car ownership have already been introduced. Chlond and Ottmann describe in *The Mobility Behaviour of Single Parents and their Activities* (2007) the mobility situation in Germany. One fifth of German family households are single-parent households. Car ownership changes with employment. More working single parents than average own cars, while unemployed single parents have far lower car ownership than average. Single parents without work have far lower car ownership rates. Working single parents feel time stress, but not many mobility problems. For non-working single parents the opposite is true. Most single parent families live in the urban areas and they mostly consider this easier from a mobility viewpoint.

Social Exclusion, Accessibility and Lone Parents (Titheridge 2008) describes the British situation. Forty-three per cent of single-parent families have no car, and this creates problems especially with regard to work (mostly working mothers) and bringing and getting children to and from school 'the ability to get children to and from school, nursery or childcare, whilst travelling and working long hours is key in terms of current UK policy'(Titheridge, 2008, 11).

The last group to be mentioned here is a rather distinct one. They are not carless, but have only one car in the family. In some circumstances this is obviously not enough. Here we enter the area of relative carlessness. The household has a car, but one of the adults cannot make use of that car at certain times. When the car is in use by the other adult, the first adult feels carless. This theme has not been analysed often, Richardson and Ampt (1997) are still the exception. However, in both cases some elaboration on this theme has taken place.

In rural Scotland women have different travel patterns to men; more journeys are related to family possibilities, often at off-peak times, and many chain journeys. In *Stuck in the Countryside? Women's Transport in Rural Aberdeenshire* (2010) Noack carried out interviews. All the women questioned considered rural living very car-dependent, with children being completely reliant on their parents, mostly on their mother 'to everything, they have to be driven' (Noack, 2010, 5). When there is only one car in the family it is mostly the woman who gets the car, she needs it for her more difficult mobility patterns. Problems arise when the car is needed by her partner. In fact, living without two cars seems to be rather difficult.

The same is the situation in peri-urban France, as is shown in *Entre ville et campagne, le difficile equilibre des periurbaines lointaines* (Ortar, 2008). Men travel for work to the urban regions, women work part-time, or after the second child, stop working. When they find work it is often in the neighbourhood. As there is very bad service on public transport '*La voiture est omnipresente, et la voiture fait partie du mode de vie et penser les deplacements autrement releve une vertitable gene, voire dune impossibilite, et ce d'autant plus que les transport en commun sont quasiment inexistants*' [The car is everywhere, is part of the way of life, and thinking about other ways of transport creates some embarrassment, is seen as an impossible route, the more so because public transport is nearly non-existent] (Ortar, 2008, 8). And interesting, '*L'absence d'un second vehicule est vecue comme une veritable contrainte pour le conjoint lese, complique le quotidian et limite fortement l'acces a l'emploi*' [The absence of a second car is seen as a real threat to the arrangements made, complicates daily life and strongly reduces access to employment] (Ortar, 2008, 8).

6.4   Social exclusion and transport

Until now the focus was on identifying mobility patterns of the carless households. At the end of the last Section we created a bridge to a wider field of research. We will now look at households that actually or potentially experience disadvantages arising from transport situations. These are mostly problems reaching locations or facilities.

In France, Australia and the UK especially over the last decade attention has been given to the relationship between social exclusion and transport. In Britain early this century a Social Exclusion Unit was created within the Ministry of Transport for this purpose. The FIA Foundation organised a worldwide study on this relation. Some more important think tanks in the US, such as the Brookings Institution show an interest in deprivation through transport issues. In contrast to the countries above is the situation in the more complete welfare states like Germany, the Netherlands or the Scandinavian countries. Here this theme, this relationship, seems to be hidden in a world of social welfare arrangements. There is very little explicit literature on social exclusion and transport to be found in these countries.[7]

What is the relationship between social exclusion and transport about? In *Transport and Social Exclusion; A G-7 comparison* (2003), Lucas distinguished certain trends:

---

7   The exception is Kemming and Borbach (2003), about the German situation.

- services and facilities move to locations that are difficult to reach without a car
- services and shops disappear from deprived areas
- personal restrictions and handicaps
- diminishing quality of public transport
- exposure to noise and air pollution from nearby transport.

Social exclusion from transport mostly deals with complex interactions between:

- the locations of activities
- the personal situation of households
- the disposal of transport possibilities.

Litman (2002) designed a circle of car dependence. In his vision car dependence leads to an increase in social exclusion because possibilities for travelling without a car are being diminished, resulting in higher travel costs. To cope with this, people buy a car and their car dependence grows. For Litman social exclusion is not only related to the carless households but also to the somewhat poorer and the poor households which have cars but have to pay a lot for transport in relation to their incomes.[8, 9]

Seen from a welfare state perspective *this theme is about the well-known problems of those at the bottom of modern western risk societies, worsened by the fact that in location decisions about work, health care and shopping good accessibility for other modes than the car is usually a non-issue.* Households more at the bottom of society (in income terms) have to spend huge sums on transport and these households often live nearer to urban roads, making them vulnerable for noise and air pollution problems.

The FIA asked authors from seven countries to report on the situation in their country, and to look at more than just the carless households (Lucas, Gentilli, Kemming and Borbach, Imanashi, all 2003; Litman, 2002; Orfeuil, 2004b). The FIA Foundation concluded 'the travel choices of people without cars have been gradually eroded, whilst at the same time the need to be more mobile has increased.'

On the basis of the existing and still growing literature four interrelated problems can be identified.

The first is *the relatively high transport costs* with which lower income groups and carless households are faced. All the FIA studies showed that poorer households spend a substantially higher share of their incomes on transport. When they are also faced with high rents or mortgages this can create stress. In the United States the impact of high transport costs together with high housing costs can be extreme. The two costs combined rise for example in Tampa, Florida to on average nearly

---

8   We already touched upon this problem in Section 5.3, and will elaborate further in this section.

9   See the Addendum for a paragraph on the price of driving.

60 per cent of the household incomes of the lower income groups. Responsible for a greater share of the transport cost is the need to travel long distances to work. Lower income groups 'are left behind as jobs relocate which leads to a decline in the tax base for the community'. We already noted an ethnic factor. Minorities face more social exclusion via transport (Surface Transport Project, 2003, 2005).

In Australia there is literature on so-called 'forced car ownership' where the poorest households live in locations where they cannot reach many activities without a car, public transport is either non-existent or going in the direction of the city centres. These groups feel compelled to buy a cheap car from their limited incomes, and are faced with transport costs exceeding 40 per cent of that income. But at least activities are accessible! (Currie et al., 2009; Currie, 2010; Johnson, 2007).[10]

The second problem consists of the *choices for the locations* of the lower income groups and carless households essential services and facilities of work, shopping and health care. These services tend to move outwards, out of the vicinity of the poorer residential areas, and more important, they often move to locations that are difficult to reach without cars. On this theme the French researcher Orfeuil has said; 'Basically the location of residences and amenities is more and more directed by the upper and middle class behaviour, for whom car use is not a problem.'(Orfeuil, 2004b, 12). In Ile de France, the region around Paris, 50 per cent of the commercial centres (with big supermarkets and entertainment facilities) do not have public transport services. Lucas shows in her paper on the British situation that in poorer urban neighbourhoods relatively few normal facilities and shops can be found. Essential professionals like doctors live far more in the richer neighbourhoods than in the poorer areas with lower health records.[11] Smaller shops disappear, because with smaller household sizes buying power diminishes and the shopkeepers who stay are increasingly confronted with criminality. Many shops are being taken over by ethnic minorities. In schools there a similar situation, richer people send their children to schools at greater distances. When poorer parents would like to do the same, there is no possibility of being compensated for transport costs.

Work locations can also be problematic. In the US the *Spatial Mismatch Hypothesis* has been developed which states that the poorer populations remain in the city, while employment moves with the middle classes to the suburbs, to edge cities and to edgeless cities. The urban poor have to travel to find decent jobs, have to pay rather huge transport costs (remember, there is little public transport in the

---

10 This can also be the case in Britain (Taylor et al., 2009, 55): 'Car usage plays a demonstrably important role in facilitating access to, and participation in, a wide range of key services and opportunities for low income households. This helps to account for the perhaps surprising levels of car ownership amongst low income groups, given the high financial costs associated with them'.

11 Wright (2008, 18) describes the US situation and concludes 'transportation barriers and specially transportation dependence, are preventing sizable numbers of persons from accessing the health care system'.

suburbs!) and are, when there is no car in the household, unable to reach the work offered and thus remain trapped in a circle of poverty (Sanchez, Stolz and Ma, 2003). Core factors in this spatial mismatch are the high costs for commuting (see Roberto, 2008), long travel times,[12] the loss of jobs in the immediate vicinities of their homes, and the restricted mobility possibilities of poorer households. The affordable homes for poorer households are situated at greater distances from the location of employment for lower skilled workers.[13]

The spatial mismatch is a serious problem in the US. Research by Blumenberg and Waller (2003) shows that while on average 8 per cent of American households do not own a car, for households with an income of US$20,000 or less (in 2003) this percentage increases to 33 per cent. Twenty-six per cent of white households and 49 per cent of black households do then not own a car. Seventy-five per cent of all Americans on welfare live in the bigger cities, and more than 60 per cent of all jobs are now found in the suburbs. Poor households buying cars are not seen as creditworthy by banks and have to pay via subprime lending (Kim, 2002) thus creating a huge financial burden.

There is now a lively debate on solutions, with two camps; bringing the work to the people, or bringing the people to the work. The first camp wants to revitalise older urban areas, thus creating new impulses and the second camp wants to subsidise poorer households in buying and maintaining cars. The debate is about pragmatism, sustainability and the wish to diminish car dependence (Goldberg, 2001; Blumenberg and Waller, 2003).

In the Netherlands the Nicis, institute for urban research, did a study on the mobility of low skilled workers (Cremers, Backera, Faun, 2007). The conclusion was that a problem of spatial mismatch could exist; the demand for their labour is in the view of the lower skilled workers so far away that 'matches' do not happen. Much new employment is situated near or along the Dutch highways, often difficult to reach without cars, and at a distance from the urban residential neighbourhoods.

The third problem is the *weak supply of other transport modes*. Not, or only at great cost, being able to reach all sorts of destinations is no great problem, when other transport modes are available. However, the growth in the efficiency of service providing in public transport seems to have led in a number of countries (Great Britain, France, New Zealand to name a few) to a decline in the supply of public transport, sometimes in frequency but more often in a smaller range of hours on which services are to be delivered (see Social Exclusion Unit, 2003). Most countries do not have countrywide coverage by the so called 'travel on demand 'services whereby people needing transport can for example phone an a taxi which then comes to pick them up. These 'travel on demand' systems tend

---

12   A study of Kawabata and Shen (2005) identified that in Tokyo public transport users within a half hour journey can find six times more jobs than in Boston, and ten times as much as in Los Angeles.

13   For a translation of the Spatial Mismatch Hypothesis to the European situation, see Todman (2003).

to be expensive and are thus mostly for designated target groups only, like the elderly, or the disabled.

Stokes (2002) and Bowden and Moseley (2006) studied car dependence in rural England. Thirteen per cent of rural households had no car and these households had, especially in areas where public transport was poor, great difficulty reaching medical services. Also shopping had become difficult with the closure of many smaller rural shops.[14]

This situation can be compared to the northern part of former Eastern Germany, a rural region that has a net loss of population. In *Offenticher Verkehr und demographischer Wandel; Chancen fur Nordostdeutschland* (2007) Heinze clarifies that public transport has become very weak: '*Im landlichen Raum ist der traditionelle OPNV zu einer Restgrosse geschrumpft*' [In rural areas public transport has diminished to a bare minimum] (Heinze, 2007, 3). Getting lifts has now taken over the role of public transport and Heinze asks the question 'why should not we just stop with public transport?' and argues for a change in thinking. No public transport just for youngsters, travelling to school, or for the non-car owners. Rather, better to concentrate on leisure, shopping and holiday travel for a 'senior oriented society' (Germany is growing very quickly into an older age society). In addition, organise 'travel on demand'!

In *Transport and Social Exclusion; New Policy Grounds, New Policy Options* (2003) Grieco puts the argument for a good and concise inventory of the services delivered for poorer households, including the supply of public transport, and its coverage throughout the week.[15] She is very critical of the focus on qualitative studies on these issues 'reports have relied in the recording of and analysing the responses of low income respondents on their transport experience, rather than undertaken the systematic measurement of the accessibility of key services to the citizens of low income areas or specific vulnerable categories' (Grieco, 2003, 4).

The last great problem is essentially the households, and their members, themselves. They miss opportunities because their perception of the transport and mobility reality, their 'mental map', or better stated; their 'travel horizon' is inaccurate or even simply wrong! They perceive fewer possibilities than there are in reality. Low-skilled workers have a specific search and travel behaviour towards work (Cremers, Backera, Faun, 2007) and in general the space in which daily life takes place increases with higher education levels. To present some figures from the Netherlands; people with the lowest education (only VMBO) travel daily 26.2 kilometres in 55 minutes, where people with the highest education (HBO/WO) travel daily double that distance (50.6 kilometres) in 84 minutes (MON, 2007). Note that travel time does not increase twice, these figures indicate the use of faster transport modes and greater travel on highways, by the best educated people. Less

---

14 Youngsters in rural areas with little education and without a car have extremely bad employment perspectives (Lyons et al. 2008, 7).

15 69 per cent of missed appointments for consultation services for infants were caused by transport problems (see Lyons et al., 2008, 10).

educated people have fewer cars than better educated people, and as we already saw, this holds even stronger for members of the minority groups.

In general less-educated people do not want to travel very far for their employment. This is even more so for people combining work with other tasks. Barriers and difficulties dominate their approaches (Morris, 2006). As the chances are higher that lower skilled workers do not live in the modern middle-class oriented neighbourhoods nearer to the highways, they will have trouble reaching highway locations physically, but also mentally! Attitude, mentality, motivation, coupled with no access to the only transport mode that can bring them easily to these locations make work locations near highways unavailable for at least a part of the lower educated groups.

Morris elaborates in *Travel Horizons* (2006) on this aspect. She describes 'travel horizons' as the distance or the location that people are able to travel. For most lower-educated people this distance is small. This is a function of knowledge, familiarity, trust, and fear of interchanges. Unfamiliarity leads to staying close to home. With their smaller travel horizons many people with lower incomes or without cars need to use services and facilities nearer to their homes. It seems that in the evening and at night their horizons become even smaller. From her research she concludes that the inhabitants of the poorer neighbourhoods of Manchester need easily accessible health care services and grocery shops. Work and school can be located at greater distances.

The *Social Exclusion Unit* (2003) formulated recommendations for creating standards and norms for the accessibility of services and facilities.[16] Mobility management is advised, and also the increase of government subsidies for public transport, or even broader; the creation of transport subsidies for lower income households. Important in their view is, as is the case in more OECD countries, that the social and societal costs are not considered when defining and designing transport policies.[17] The Unit also show that with the subsidies on rail transport higher incomes households are supported. The poorer households do not travel by rail over longer distances.

Interesting and important problems but these problems get only minor attention in the political and societal debates. The reasons for this lack of attention on these problems can probably be found in three elements, the last being somewhat country specific.

The first reason is the arrangements of the welfare states. Orfeuil (2004b) signals this, and Kemming and Borbach show that these problems are, in welfare states, part of the greater societal question of paying for the maintenance of solidarity. Themes like employment, work security, functioning of the labour markets and

---

16   On this issue, see Solomon and Titheridge (2009).

17   For example on the Netherlands Boon, Geurs and Van Wee conclude in *Sociale effecten van verkeer, een overzicht* (2003, 18); 'social effects of traffic (and meant is social cohesion, social inequality, health impacts) get far less attention in research and policy than economic and ecological effects'.

health care are seen as more important. At the same time in the complete welfare states arrangements for the lower incomes exist, by which the problems do not get the same harsh character as in most Anglo-Saxon countries.[18] For example; the Netherlands has an integrated policy for the poorest neighbourhoods, in which, implicitly, all societal daily life questions can be covered.

A second reason is the system of 'travel on demand', or, as it is called in the Netherlands, the target group policy for transport (*doelgroepvervoer*). Elements of this system are also used in other countries, or are now being set up. Travel on demand creates an extra alternative for carless households, or for lower income groups, as long as they fit certain criteria. For students there is a cheap travel card, for the unemployed who need to start working again there is a subsidy on travel costs, and there are different subsidies for the elderly and the handicapped.[19] This '*doelgroepenvervoer*' costs more than one billion euros per year, and is as expensive as the subsidies on regular public transport but from evaluations it can be noticed that the target groups like this system, because it offers transport by cars when they need it.

The third and last reason is more country specific. The Netherlands is densely populated and the bicycle can be used everywhere. This means the Netherlands has two advantages compared to less densely populated and more hilly or mountainous countries.

As bicycles are ubiquitous in the Netherlands, as in Denmark, those two countries have an extra transport mode, available for almost all at low cost, and to be used on shorter journeys when there is not much luggage to carry. However, there remain problems for the elderly, the disabled and, because they do not cycle often, ethnic minorities.

High population densities make well-equipped public transport possible. This advantage in the Netherlands is shared with most big and bigger cities, and with the more densely populated urban regions of Western Europe. An inventory of the quality of Dutch public transport was done in 2007 by the *Kennisplatform Verkeer en Vervoer*. The figures presented show that between 2000 and 2004 the number of serviced kilometres (length and frequency) did diminish with 1 per cent, but increased between 2004 and 2007 with 2.4 per cent. The number of stops has diminished with 10 per cent, the frequency has slightly increased, and the amount of kilometres travelled by the passenger did increase slightly. The overall picture is a very small growth, and a trend to concentration of the services at the busiest times. Services early in the morning, later in the evenings or on Sundays are sometimes lost.

To conclude, there seems to be a spectrum in countries expecting social exclusion from transport. On the one end of the spectrum we find not densely populated countries, with insignificant public transport, and without the full

---

18   Here a relationship can be seen with the paper by Coutard, Dupuy and Fol, described in Section 3.5.

19   An overview is offered in KiM,2009d,43.

range of subsidies from the state. In these countries, for example the US, the UK or New Zealand and the peri-urban areas of France, social exclusion by means of transport should be an important issue on the societal agenda. On the other end of the spectrum, in densely populated countries, with rather well-developed public transport, and with a complete welfare state, like Germany, the Netherlands or Denmark it can be expected that problems are on a smaller scale.[20] In the Netherlands signs of social exclusion trough transport are seen in the more rural regions. Harms (2008, 195) showed that rural households without cars can only get 'travel by demand' at high individual cost.

For the future of Western Europe it will be important how the following factors will interrelate for the low income households and for the carless households:

- greater travel distances to get to services and facilities
- greater commuting distances
- working locations are more difficult to be reached without a car
- employment possibilities are beyond the 'travel horizons'
- the costs for driving will increase, as seen in Chapter 5
- fuel prices may be less expensive through energy efficiency.

There is a chance that driving will become more expensive for poorer households. Unless government policies are prepared the four problems mentioned will be with us to stay. More research on social exclusion, household budgets and accessibility of reasonably-priced transport is a necessity!

6.5 Frequent car use and car dependence at a societal and cultural level

In this section we will broaden the spectrum of issues once again. Frequent car use changes something in modern western risk societies. In Chapter 3 we tried to clarify how, because of the convenience and the facilities offered, the car could become a basic asset in modern societies. Cars have made the geographical spread of locations for important activities possible. 'Vicinity' or 'nearness' is, outside the cities, no longer the standard in daily life. Even in cities there is sometimes a need to travel longer distances to be able to carry out all the wished, expected and necessary activities of daily life. In this section we will look at the social changes connected with frequent car use. We will talk about a double relationship; sometimes car use leads to social change, and sometimes it is the other way around. We will describe the rise of the edge cities, of the edgeless cities, of mobility in the newest modern neighbourhood in the Netherlands. And we will describe hypermobility, giving rise to permanent moves and to daily fluctuations

---

20 It would be interesting to know the position of the other Scandinavian countries; Norway, Sweden and Finland are welfare states, but have low population densities. Little is known about their social exclusion through transport problems.

in population composition of locations. Finally we will focus on some cultural orientations.

Community light

Social cohesion is a concept used for the connection between people. The concept is often used for territorial bonds but this is not obligatory. Territorially indicated, the concept defines the measure of commitment between people, in a finite space. Many authors (like Sennett, Bauman, Boutellier and Durodie) state that social cohesion, described as above, has decreased. Other authors (Castells, Elchardus) do not share this view and explain that social cohesion has been nested at a higher spatial scale. In their view we no longer feel connected to our village, our city, our region, but more to our country, or even to the world at large. For these authors the distance for cohesion has grown tremendously. In a third vision the 'spatiality' expires; social cohesion can be found in non-spatial entities, such as groups and networks created through digital activities (e.g. Hyves and Facebook memberships).

In the US, over recent decades a great loss of function can be seen in the city centres in favour of the edge cities (Garreau, 1991), the built-up areas situated near highway crossings where office developments form the basis behind the creation of urban networks of work locations, shopping malls and housing in low density. These edge cities were followed up in the so-called edgeless cities. In an overview article Lang (2003) paints the developments of these cities. These are forms of 'sprawling office developments' in a spatial order without focus. Later, housing also appears in even lower densities than in the edge cities. Especially in the south of the US there are many edgeless cities. Thirty-seven per cent of office employment is now found in edgeless cities, for the edge cities and for downtown areas the figures are 21 per cent and 38 per cent. The expectation is that, contrary to the edge cities, the edgeless cities will have no growth in density, 'introducing a city-like street network into places that are now built around automobiles will be impossible' (Lang, 2003, 11).

In the Netherlands and more widely in Western Europe the newest housing developments come near to these American developments. These new housing developments, called Vinex areas in the Netherlands[21] are situated nearer to highways than to the cities where they form part of. The inhabitants of the areas orient themselves on the locations within easy reach via these highways. Reijndorp et al. introduced the first analysis in *Buitenwijk* (1998). In their view the inhabitants like their house, are indifferent to their neighbourhood but like the location of their neighbourhood on the Dutch map; everything is within easy reach. The possibilities that the highway offers in terms of reaching other places is far more important than their orientation towards the city of which they form the utmost outlying part (sometimes over 10 kilometres from the city centre). Social life for these inhabitants

---

21   These Vinex – areas represent 15 per cent of the Dutch housing stock.

is not concentrated in their new neighbourhoods. This neighbourhood is just the first step in the wider areas where they life and create their activity spaces. With growing education the 'zone of interest' becomes greater.

The Vinex demography is more heterogeneous than expected. In these neighbourhoods live many singles, young people, and middle-class families but there is not a complete domination by families (Atzema and Hooimeyer, 2000). There are few older people and the less well-educated are missing. Car use is rather high in these neighbourhoods and in general the residents travel more kilometres than on average, and the difference is growing; on average Vinex residents drive 4 kilometres further than average, some 12 per cent more than average (Snellen, Hilbers and Hendriks, RPB, 2005, 66). This effect also grows over distance – each kilometre further from the city centre leads to 130 metres extra car travel each day per person (Snellen, Hilbers, and Hendriks, 2005, 54). Full-time workers are the greatest car users, but they are followed closely by part time working higher educated people, a description of the 'rushing around mothers'. It is interesting to note that children from these neighbourhoods walk and cycle significantly less than their age group in all other locations (Snellen, Hilbers and Hendriks, 2005, 52). They are usually driven everywhere.

Some Vinex residents always use their cars and are not concerned about the characteristics of the environment, while a smaller part will consider alternative transport modes. The most sensitive are students, better-educated people and task combiners (Hilbers, Snellen, Hendriks, 2005, 59).

Most newer neighbourhoods are not inhabited as collective shared environments and living worlds and ties between the residents are few, although there is a difference between the 'pioneer' phase (with greater solidarity) and the 'built up' phase, where there is no emotional bond with the neighbourhood (Heijmans, 2007). The neighbourhood does not function as a permanent place to live – it feels more like a half-way house. Nio (2006) constructs the bond that belongs with this emotion; loose, pragmatic, functional. Others describe this bond as weak, not profound, or nicer, as 'community light' (see Lupi, 2005, and Lupi and Musterd, 2004). More scientifically 'most residents want to identify with their neighbours in a way that is characterised by a certain distance in combination with easy, flexible and shallow contacts' (Hortulanus and Machielse, 2001, 12).

From the neighbourhoods we move to society at large. The way the comprehensive role of the car is valued in modern societies depends on the vision of authors. First in *Hypermobility; a challenge to governance* (Adams 1999, 2005) he paints a clear picture. Hypermobility is, in his eyes, unlimited, unbounded physical mobility. Characteristic for this form of mobility are:

- spread built up areas
- polarisation between income groups
- dangerous and widespread criminality
- hostile towards children
- fat and less fit

- not varied culturally
- anonymous and more suspicious
- less democratic.[22]

The cohesion in such a society will be low. Adams asks: what would be the distinctive characteristic of a policy aimed at an increase of car dependence? His answer is 'it would be a package of measures designed to encourage people to move out of town and spread themselves about in densities that were too low to be served by public transport' (Adams, 2005, 9). To effectively block a further development of car dependence 'governments would seek to restrict traffic in areas where its growth is fastest – not in congested urban areas, where it has already stopped, but in the suburbs and beyond' (Adams, 2005, 9). He noticed that politicians are afraid of acting this way. And so a complex situation remains; people like to drive unlimited, but they do not want to live in a world that is the result of this unlimited driving.

Sager continues this line of thinking in *Footloose and Forecast-free: Hypermobility and the Planning of Society* (2005). His statement is that an extreme mobility will undermine the capacity of planners and policy makers to predict, and also the capacity to create with these predictions in mind a well-considered and well-balanced societal system. In his vision there is a narrow relationship between unbounded mobility and society being unpredictable. And there is a great tension between hypermobility and 'the Enlightenment project of self-determination through knowledge-based social decisions' (Sager, 2005, 4). The decreasing importance of distance lessens the chances of keeping territorial communities alive. Mobility creates opportunities to break out of the exigencies and rituals of territorially bound communities. 'High mobility means victory in the struggle against distance' and 'revealed hypermobility dissolves community' (Sager, 2005, 9); with complete free mobility everyone can follow his or her own optimum, and organisation of the forms of collectivity becomes impossible. Sager worries a lot about the consequences for the process of collective will when everybody can easily be mobile. The possibility to predict and to create collective acceptable approaches to societal problems will disappear, and in his view 'this is fatal to the modern idea of man as the master of his own destiny' (Sager, 2005, 16).

Both Sager and Adams are the exponents of a movement that sees an increase in mobility (by which they mean car mobility and air traffic) result in uncontrolled, unrestrained, non-democratic, spread out and norm-less societies. In their[23]

---

22   These characteristics are evaluated on their relevance in traffic terms in RIVM (Geurs and Van Wee, 2000).

23   The literature has remarkably few descriptions of the positive value of increasing car dependence for social cohesion. Exceptions are: RAC Foundation (2009), Lucas, Grosvenor and Simpson (2001) and the vision from the American think tank Reason Foundation; Dunn (1998), Lomansky (1995) and Staley (2006). See also Section 7.5.

opinion hypermobility kills social cohesion, and causes irreparable damage to the idea of a responsive and responsible society.

Hypermobility and fluctuating networks

However, there is certainly some value in car dependence for certain aspects of social cohesion; the car makes the set-up, the maintenance and the extension of varied and geographically spread friends and family networks far easier. It is very difficult to visit friends who live 70 kilometres away in the evening without a car. And it is also not very easy to give help on a more-or-less permanent basis to older relatives or friends living at some distance, without a car. Public transport takes longer, is not available at night, and does not fit into the busy schedules of modern people. Volunteer help is possible because it can be delivered by car when needed. We do not live in an 'easy-going' world! Older people complain and see their worlds shrinking when they have to give up driving (see Davey, 2004; Department of Infrastructure Victoria, 2007).

Basically the car plays a paradoxical role; the car creates the build-up and maintenance of contact networks over greater distances, and at greater geographical scales, while at the same time killing the need to invest and to keep investing in permanent contacts on the small action radius of the own neighbourhood, own village or city.

For the middle and higher classes in modern western risk societies this is primarily a neutral statement. But this changes when looking at less-educated people, who live, in essence, more local lives. Their action radius is far more often limited to where they were born and raised and their travel horizons are far smaller (Morris, 2006 and Cremers, Backera and Faun, 2007). This is also the situation for most non-car owning households, with the exception of a group of cosmopolitans (see Section 3.1). A further increase in car dependence will lead to diminishing social cohesion at the neighbourhood level, which is for lower-educated and low income households the most relevant scale. This development will be partly compensated by the build-up of lighter forms of social cohesion (like 'community light') at a higher level, but something for the lower educated will certainly get lost.[24]

For most middle and higher class households geographically spread (and spreading!) networks of friends are their primary source of social cohesion. Community life in their own neighbourhoods or municipalities is for them of lesser importance. Axhausen, Urry and Larsen elaborate on this issue in *The Network Society and the Networked Traveller* (2009). Every person has his networks, and

---

24   It is an open question whether it has not already been lost. Politicians especially play 'lip service' to maintaining social cohesion in the poorer neighbourhoods of our societies. Mostly they seem to do this because their fear of inter-ethnic problems. It is questionable whether permanent cohesive lower-class neighbourhoods are still possible under the core arrangements of our modern risk societies, that reward flexibility far more than maintaining community values.

travels around within this network. Network travellers mostly do not share close bonds with their neighbours. They no longer live local lives. There are some neighbourhood contacts, but they do not dominate 'for the bulk of the residents the immediate environment around their residence is populated by strangers'. And, connecting to the vision of Durodie in Section 2.4, this freedom from social control of the neighbours comes at a cost in times of crises.

Schokker and Peters focus on hypermobility from a completely different angle than Adams and Sager. In *Hypermobielen* (2006) they look at what hypermobilists are actually doing. Hypermobility from them is an expression of hypermodernism, and that is just modernism to its utmost consequences. They see a 'strong individualistic life style coming up in a global functioning economy'. With that life comes 'in increasing levels; an increase to the "now", the "moment", experiencing "real time" and no acceptance of delays or postponement whatsoever' (Lipovetsky)[25] (Schokker and Peters, 2006, 6). Time management in such a society is, for every individual, a personal task. The car has here a specific role. As the car is available, expectations regarding time management and combining activities are rising. Friends expect you to be able to combine activities and work within tight time schedules because you have a flexible tool, able to solve mobility puzzles. 'Hypermobiles' are (assisted by mobile apparatus, not only the car but also mobile phones, smartphones etc.,) experienced in solving these puzzles and changing schedules when necessary. Mobility for them is not just going from A to B[26] but a perpetual search for the cleverest routes, necessary to combine activities within shifting time frames. Obviously these 'hypermobiles' consider themselves car dependent!

Hypermobility also gives rise to another phenomenon. The composition of the population at a specific location changes during the day. Nuvolati elaborates on this element in *Resident and Non-Resident Populations; Quality of Life, Mobility and Time Policies* (2003). The idea that 'the space in which the community is located is daily or seasonal populated by different groups coming from different places and staying for different quantities of time during the day' (Nuvolati, 2003, 69) is central in his view. The social cohesion changes with the groups present on the location at any time and there can be mutual conflicts between the groups present at the same time. Conflicts can arise over the interpretation of the physical space; is it the activity space for the businessmen, or the daily space for the permanent inhabitants? Conflicts also arise over accessibility; is easy accessibility always prevalent or should speed limits be used to make neighbourhoods better to live in? Conflicts about local taxes; the permanent inhabitants usually pay for facilities used by temporarily active workers or shoppers, who will mostly have a 'free lunch'. And more basically; what constitutes the 'community' at a certain location, when the composition of the population changes hourly? And what is the standard for social cohesion; the vision of only the permanent inhabitants, or also the vision of all the people who inhabit the location over a day?

---

25 The French philosopher Lipovetsky paints in *Les Temps Hypermodernes* (2004) the rise of a society of instant action, cherishing to 'now' and the experience of 'real time'.

26 As is still the case with public transport.

It seems wise to differentiate in at least four groups; inhabitants, commuters, business men and users of the location, like shoppers and sportspeople. Neighbourhoods, especially working areas and city centres, are in constant flux, the composition of their population changing all the time. During the working day employees, who live elsewhere, outnumber the permanent inhabitants, while at weekends many permanent inhabitants go away and are replaced by leisure searchers and shoppers.

In *In Perpetual Motion* (2005) Zandvliet makes a differentiation between inhabitants and the regular temporary populace. Some cities have better economic results than could be expected from their inhabitants. Here the regular temporary populace, its added value and its spending, make the difference (see also Zandvliet, Dijst, and Bertolini, 2002).

To conclude; social cohesion has become a difficult concept. The car creates the possibility to withdraw from the social and cultural rules of your neighbourhood. On the other hand the car can help creating social cohesion at higher geographical scales. While hypermobiles try to reach this cohesion with their ultra-flexible behaviour, those with lower incomes, the lower-educated and most carless households need cohesion at neighbourhood level. But regarding this neighbourhood level the question has risen; social cohesion, for whom? The composition of the population changes all the time, due to the flexibilities that the car helped to create.

Cultural orientations

In this section we will look at some cultural aspects of car use. Culturally a division around the positive and negative aspects of the car can be seen. With Freund and Martin, cited by Soron (2008, 184) it can be seen that the advantages of cars and car use are to be found in the domain of the individual possibilities. Ownership of a car broadens the spectrum of possibilities in space and time, leads to convenience and facilitates new activities such as hang gliding.[27] The disadvantages are more-or-less completely to be found in the domain of the commonalities; the car kills the need to invest in social cohesion in neighbourhoods, disintegrates public spaces, creates environmental problems, and creates extra difficulties for non-users to take opportunities, their living conditions deteriorating as a result of ubiquitous use of the car.

The growing car dependence of our modern western risk societies has, unintentionally, got the characteristics of an extreme liberal[28] project. The car fits perfectly in a system to 'let everybody be happy in its own way'. Negative effects of car use are mitigated (traffic safety, nature, landscape, air pollution, noise), redeemed (special travel on demand for the elderly), or hardly noticed (loss of playing spaces for children, loss of quality of public spaces). An important question

---

27   Looking at the luggage involved, and looking at the specificities of the start locations it seems impossible to go hang-gliding by public transport!

28   Liberal in the sense of Margaret Thatcher's 'there is no such thing as society'!

is whether with the battle against global warming, and broader with the objectives of sustainability to be reached, this extreme liberal project has not reached its frontiers. Division of scarcities asks for another political energy. However, it has to be acknowledged that, even as car mobility is the only still growing emitter of CO_2, there is a hardly an important movement to diminish car mobility and car dependence in our western risk societies.

At the individual level a few cultural changes can be noticed. In most countries can be seen that the more a transport system is car oriented, the less adults are walking or biking at a regular basis (Garling and Loukopoulos, 2005; see also Section 3.2; Freund and Martin, 2004, 276). In the United States most new neighbourhoods are now designed without cycling and walking paths. Freund and Martin show in *Walking and Motoring: fitness and the social organisation of movement* (2004) the decline of time available for spontaneous, physical, non-disciplined, activity;[29] 'to walk is to contest the standard space–time usage' (Freund and Martin, 2004, 277). A car-defined spatial organisation of activities is introduced, with much spread in locations, and with a need to drive to get along. The effect of far too little obvious, natural and spontaneous movement can be seen in the increase of obesity.

In OECD countries the obvious movement, still to be seen in developing countries, has been replaced by a movement on choice. Moving your body is disconnected from normal daily life and given a defined place in jogging hours and fitness minutes. People now travel by car to use walking machines in fitness centres!

Because this whole system around moving is built on choice, and not on need, many people withdraw and can develop in the direction of motionlessness.

Freund and Martin also make a connection between the car, and another cause for obesity, the food we eat. In *Fast cars, Fast food; Hyperconsumption and its health and environmental consequences* they show how both elements fit in the arrangements of daily life 'possession of a car in the US is a necessity, fast food for time constrained people a reality' (Freund and Martin, 2008, 4). Obesity problems in the US are concentrated in sprawling districts, and in city centres in the US it is rather difficult to get good food without a car (the 'food deserts' see Larsen and Gilliland, 2008).

Much research has been done on the geographical spread of obesity problems in the US (Ewing et al., 2003, 50). In 2005 *TRB* together with *the Institute of Medicine* in 2005 carried out a comprehensive study on the relationship between transport, the physical environment and health aspects. In *Does the Built Environment influence Physical Activity?* the researchers arrive at careful and balanced conclusions 'the available evidence shows an association between the built environment and physical activity. However, few studies capable of demonstrating a causal relationship have been conducted, and evidence in supporting such a relationship is currently sparse'

---

29 Jogging and fitness are not spontaneous, but planned and disciplined activities, completely fitting in car cultures!

(TRB and Institute of Medicine, 2005). The study concludes that people living near highways have more health problems connected with noise and air pollution.

Car driving can lead to stress, which in turn can create risks, because driving is not an isolated activity and is carried out in often complex traffic situations, and under time pressure. However, for most car drivers, stress is not an issue.

Stress can have many faces

First is the stress and uncertainty about the cost of driving. Low income households sometimes question how long they will be able to keep driving, looking at the money available for transport in their budgets. It is to be expected that this stress will increase over the next decade.

Second is the stress related to driving itself; to complex traffic situations, to journeys that take longer than expected, thus raising problems with appointments. The unreliability of travel times can be a source of stress. Stutzer and Frey looked in *Stress that does not pay; the commuting paradox* (2004) at the situation in Germany, where people with long commute times are less satisfied with their lives. The authors suggest that one major reason is under-estimating the capacity to adjust to long working days. The advantages of living in rural areas do not compensate for the loss of leisure time and the necessity to travel each day the same long distance every day.

Stress can have strange faces. We already noticed (in Section 3.4) road rage. Drivers can be made to feel insecure by these 'dangers on the roads' and there is driving anxiety. In the Netherlands some half a million car drivers are anxious about driving. There is little literature on anxiety and stress in traffic.[30] However, there is a vast literature on traffic safety issues. This situation calls for further analysis. One possible explanation is that car drivers convert their feelings of anxiety and stress into a preoccupation with safety in and around their cars. They objectivise their anxieties and that has probably to do with their need to drive. You cannot think all the time about necessities!

6.6   Scarcity in driving opportunities

This is the first of the two *capita selecta*. In these *capita selecta* my own views are presented, somewhat more strongly than in the other parts of this chapter. We leave the boundaries of academia a little.

Here we look at congestion in the Netherlands. Congestion is a traffic problem with a middle class bias. Commuting over longer distances is an activity more related to the arrangements of middle and upper class life. As we have seen the lower classes stay within their own neighbourhoods more, and travel smaller

---

30   On driving anxiety, see Van den Berg, *Omgaan met rij-angs*, 2005. And more broadly, Mesken (2006) and Levelt (2002).

distances. In the Netherlands car drivers cannot, at all times, travel to all locations in the time they had counted on or expected. In essence that is the problem of congestion. Standing still in your car is, during rush hours in the Randstad area, normal practice. On certain roads at certain hours the road space is just too small to accommodate demand.

In essence, two approaches can be taken – creating extra road space, or changing demand. First creating road space: This is an expensive solution for a periodic problem. Governments will not usually build roads or extra lanes just for rush hours, which are only 20 per cent of the hours in a day. However, some extra capacity is wise, and efficient. Extra road capacity can also be found by a better organisation of the traffic flows. Here we enter the domain of road operations. Modern traffic management, in relation to travel information on the spot can increase road capacity.

Changing demand is another story. The fact that roads are completely full at certain hours has to do with the circumstance that our western societies are all organised in more-or-less the same time frames. Working generally starts at eight in the morning, and ends at six in the evening, so it is understandable that there will be heavy traffic at related times. Creating better societal arrangements is a necessity – working from home, flexible working hours, more differentiation during the day, this should be core solutions. However, that being too difficult to organise for employers, governments and shopkeepers, the focus is on complaining.

But pragmatic elements are available. Mobility management can be useful (see *Taskforce Mobiliteitsmanagement,* 2008) and charging for the use of the road, when road space is scarce would be helpful. However; politicians have difficulty explaining the wisdom of this solution to their voters, one reason for a constant failure in the national political sphere. To quote Stopher and Fitzgerald in *Managing Congestion – Are we willing to pay the price?* (2008, 10) 'Simply stated, because roads in most places are free at the point at which they are used, there is little or no financial incentive for car drivers not to overuse them.'

The congestion problem can certainly be solved but as long as the different responsible stakeholders are not willing to fundamentally change their arrangements and attitudes, and as long as road space is not seen as just a scarce commodity, with a price, we will keep talking about this problem. And that is what happens; the congestion is the only problem with social and cultural aspects that gets attention in the media, in the political and societal debate, showing that mobility policy is essentially middle-class oriented.[31] More attention to the other problems mentioned in this chapter in relation to this, without fundamental changes, unsolvable problem would be wise.

In the Netherlands the focus in mobility policy is on '*bereikbaarheid*', to be translated as accessibility. It is interesting to see how this objective is clarified. Accessibility is concerned with being able to reach destinations within a certain

---

31   Understandable while in modern western risk societies this middle-class is huge, and functions as the backbone for society!

time, and with some reliability. Note that this is an objective for households (with one or two cars, and with no big budget problems) that have no problem reaching their destinations – these destinations are perfectly accessible for them. They only want to be able to go faster and in more reliable time frames to these destinations. This means that increasing the speed of travel, and diminishing insecurity are the core objectives of the Dutch *bereikbaarheid*/accessibility policy. Under this policy, somewhat strange looking at its name, there is no attention to the real accessibility problems faced by the lower income and the carless households. These households are the ones that cannot reach facilities, services, or more widely their destinations, with the same relative ease of the middle classes!

A part of the problem of congestion is geographical in origin, the working areas, the housing areas, are spread out. The relationship between spatial forms and spatial planning and car traffic and car mobility is a long standing research theme, especially for academic researchers. Which form of an urban region generates more mobility, and how can we order space in such a way that less car traffic and less congestion are the result? These questions are investigated over and over again. However, most analytical research shows a rather weak relationship between spatial arrangements and mobility. As Snellen concludes in her thesis *Urban Form and Activity-Travel Patterns* 'the effect of urban form on travel behaviour is rather limited' (Snellen, 2001, 125). And Schwanen, Dijst and Dieleman arrive in *Policies on Urban Form and their Impact on Travel: The Netherlands Experience* (2004, 594) at the conclusion that the significance of urban form to the behaviour of commuters should not be overstated. 'Limited, but of some significance' is the verdict.

In an overview article, *Smart Growth and the Transportation – Land Use Connection: What does the Research Tell Us?* (2005), Handy offers a few relativistic conclusions. Constructing new roads or new carriageways will create some extra sprawl, and will lessen congestion somewhat, but not very much; built up areas already cover much land. Investing in light railways will create somewhat higher densities, and proper spatial planning could diminish car use somewhat. No big changes to be noticed.

It becomes clear that the greatest explanation for commuting and for congestion arise from the individual attitudes and motives, as discussed in Chapter 3. Physical planning strategies are far less important.

The question here is: if the attitudes and motives are far more important in explaining car use, than the relationship between housing densities, spatial planning and car mobility, why is so much research effort put into just these issues, and relatively little into social and cultural orientations and problems?

Here again we see a focus on themes that are not the most important, when one tries to understand car mobility, its sources, problems, added values and perspectives. An explanation could be that this relationship between land use, physical space and mobility is a relative safe research topic, and is not so 'value laden' or political dangerous as the themes on the vulnerability of societal groups, or social exclusion.

6.7 Physical mobility and virtual mobility

It can be expected that, with some 80 per cent plus of all kilometres travelled by car, and an increase in car dependence to 50 per cent:

- it will be more difficult for carless households to participate fully in modern society, missing chances to join in activities at locations organised around the car;
- transport will be more expensive for low income households, in relation to their budgets;
- social cohesion at the neighbourhood level will diminish, giving rise to a spectrum of anxieties;
- social cohesion on higher geographical scales will increase, but not permanently, more as flexible and fluctuating arrangements;
- opportunities to drive will be scarcer, and congestion will still grow somewhat.

In this last *capita selectum* we will look at the way virtual mobility can change the above framing. Information and communication technology (ICT) can be seen as important in changing frames in society. Sometimes it is forecasted that the virtual mobility, facilitated by ICT, will diminish the need for physical mobility. People do not have to leave their homes to communicate, they can do this online. A complete network of contacts can be organised and maintained via computer, Blackberry™ and mobile phone.[32] But will virtual mobility really take over a part of the physical mobility?

In *Social Implications of Mobile Telephony; The Rise of Personal Communication Society* (2008) Campbell and Park conclude that mobile phones make a more personal relationship to time and space possible 'individuals reconstruct the meaning of time and space for personal purposes as they rely on mobile telephony rather than set places and set times in their efforts to coordinate with others'. Mobile communication personalises public spaces. In their view we are becoming a personal communication society. The function of the car in such a society has to be decided. Thulin and Vilhelmson's *(*2007) research on young adults in Sweden indicated that the most mobile active persons were also the most communication-oriented. Mobile communication strengthens face-to-face contacts and the other way around!

Ling and Haddon elaborate in *Mobile telephony: mobility and the coordination of everyday life* (2001) on the role of the mobile phone in micro coordination. They indicate that mobile phones create a capacity to access extra information while driving, thus creating possibilities for more efficient use of the transport system. Coordination via mobile phones saves journeys, is their conclusion. This seems to

---

32   And the mobile phone offers the possibility to receive online travel information during your car trip.

be in line with later studies, where consensus is growing that mobile phones can deliver a small, but positive, contribution to the efficient use of transport systems.[33]

A decade ago the first articles on the relationship between virtual mobility and factual physical mobility started to appear. At that time there was a 'substitute-vision' (virtual mobility will replace physical mobility) and a 'complement-vision' (virtual mobility and physical mobility are complementary). The 'complement-vision' describes the actual situation better; physical mobility has not, or only slightly, diminished, while virtual mobility grew quickly. Working from home is still rather marginal, with 6 per cent of the workforce on a more or less permanent basis. The great jump has not been made.[34] Which driving forces are responsible for preventing substantial changes?

Three probable explanations can be offered; habit, the wish to meet in person, and active resistance by employers.

To start with the last explanation: When employees work at home employers cannot control whether they are working. Working at home is seen as problematic by many employers, especially by operational managers. Although the management literature sometimes suggests otherwise, the basic management style in offices and factories is still 'command and control', especially at the lower management levels. With their employees working at home, managers have to steer on results, on content, and not on behaviour in the office. Many managers are not able to steer that way, are not used to manage professionals. Cap Gemini (2009) describes this in its report *Trends in Mobiliteit*. Although it should not matter what employees do all the day – when they achieve the necessary results and are available to talk to and email with colleagues, that should be enough – this new working structure is blocked by old fashioned control thinking.[35]

The wish to meet face-to-face plays a role in difficult decisions and in shopping and leisure. Virtual meeting remains something other than actual meeting, and habit remains, as we clarified in Section 3.6, an important explanation, especially for the generation over 30, who did not grow up with ICT. Many ICT possibilities seem to be used in practice for the first time when employers see physical mobility to work no longer as obvious, and start considering organising work in a 'mobility poor' way (Cap Gemini, 2009, 25).

Possibly we will arrive at a tipping point (Gladwell, 2000) when the first generation that grew up with ICT reaches adulthood and starts driving. For this generation the complete use of the ICT spectrum is obvious. This generation created virtual networks, and lives in them. The mobile youth culture creates permanent communication with friends, who are being seen in real life or on the web and with

---

33   However, Van Wee and Chorus (2009) show in their overview that we know far more about the effects of ICT on travel behaviour than of the effects of ICT on accessibility.

34   On how such a jump could appear, see Cairn et al. (2004, 345). With their strategy a reduction in car commutes of 12 per cent (already in 2004!) could be reached.

35   It is clear that working at home would not be possible in all jobs, for example in health care and education it would be difficult to implement.

whom experiences are shared. For youngsters it does not seem so necessary to travel to friends, one can see them just as well online. They are at home, but connected and spend each year more time behind the computer each year (Funk, 2009).

Manderscheid takes up this point in a lecture called *Automobile Subjekte* (2012). In her vision the dominance of the car, the automobile subject, will lose its hegemony and the 'creative nomad equipped with various mobile technologies and connected to Internet' will offer a new frame.

Cap Gemini (2009) signals that car ownership by their young professionals is decreasing. Their new employees now prefer an annual ticket for public transport over a company car. Shell presented in 2009 its *PKW-Szenarien bis 2030 for Germany*. In the 18–30 age group Shell sees car mobility diminishing especially under men. Shell also sees a more cultural reason[36] '*Viele dieser junge Menschen wohnen noch zu Hause oder in Gemeinschaft mit anderen jungen Leuten, wo sich nicht jeder einen eigenen Pkw leisten kann, eventual aber auch nicht jeder einen eigenen Pkw benotigt oder gar haben will (sogenannte ecoboomers)*' [Many youngsters still live at home or with other young people, and they cannot pay for their own car, or are not willing to have a car (the ecoboomers)] (Shell, 2009, 21). And Van der Zwart (2007) states that, where youngsters can be in and with the whole world, they have a demand for a clear local identity.

The Social Research Institute SCP did a study in 2009 on ICT use of young people with the title *Nieuwe Links in het gezin*. Young people are Internet focussed on entertainment and communication; 'youngsters are the authors of their own life. The social network is constructed, and can be visited all the time'

We have to conclude that we still know too little about the physical mobility of young people between 12 and 25. There are chances that, with their direct contacts via the Internet, MSN, mobile phones, webcams, they will tend to more selectivity about their physical mobility – the first signs are there. To finish a quote from the Danish researcher Laessoe in *The Need for Mobility* (2001): 'With the advent of cyberspace the issue is no longer a simple one, of faster communication across longer distances, but much rather one of communication and experience in a 'space without a place…this situation provides new opportunities for social interaction and intensity, while also inserting a protective filter that allows for and enlarges our fear of attachment. Since it also reduces time for other activities, it could imply that our need for mobility would be allowed to develop further, while actually reducing our material transportation requirements' (Laessoe, 2001, 104).

6.8 Instead of conclusions

The problems mentioned is this chapter have at least one element in common: they do not receive much attention in the societal and political debates related to car use. These are problems signalled by academia, and by the concerned households

---

36 Clarified for the age group 25–30 years.

themselves and they tend to be forgotten on a societal level, or to be seen as not very problematic. Only one problem is given any real attention, the congestion on some roads at some times.

Perhaps the problems mentioned here have two other elements in common. The first could be that their present magnitude is somewhat unclear. How great are the problems mentioned here? It looks that here is an element of appreciation involved. The value attached to these problems differs with the normative viewpoint of the beholder. Take the loss of social cohesion on neighbourhood level. This can be seen as problematic, losing some fine elements in the intricate woven tapestry of society. But it can also be seen as just old fashioned thinking by people who still do not want to see that 'there is no such thing as society'.

The second is that almost all problems mentioned are problems of the lower classes, the lower educated, the households on the wrong side of society's welfare. This is a small exaggeration, taking into account the fact that at least a part of the carless households chooses a carless existence but this remains a relatively small part.

What can be stated fairly certain is that with further car dependence without mitigating measures the position of carless households will be further marginalised. What also can be stated fairly certain is that with higher prices for fuels and for vehicles the position of the poorer households will become more difficult. An even bigger part of their budgets will have to be spent on mobility. It will, with these two trends combined, lead to more difficulties in access to shops, health care, work and services for at least a part of the population in modern western societies.

There are two extra conclusions to be drawn. The first is the bias in mobility policy towards the only middle class problem with social and cultural aspects, the congestion. That this problem is the only one receiving real political attention indicates that *mobility policy in our western risk societies is essentially policy for the middle classes*. Their risks are taken very seriously by decision makers!

The second is the bias in mobility research. The situation whereby so much research funding is spent on the already well-known relationship between spatial planning, the built up environment and mobility indicates that researchers and research investors tend to focus on neutral research questions. Asking really difficult questions about motives, attitudes, anxieties and fears related to car use seems to be too challenging.

Chapter 7
Scenarios for the Governance of Car Mobility

7.1 The Governance Question of Car Mobility

In Part 1 we have seen:

- a great share of the car in mobility (80 per cent of travel kilometres),
- a trend towards an even higher share, and
- a development towards greater car dependence in the modern risk societies of Western Europe.

In the first two chapters of Part 2 we looked at future problems on car mobility.

In Chapter 5 it became clear that towards 2030 car drivers will probably be faced with:

a. higher fuel prices
b. lower fuel use per kilometre
c. higher car purchase costs for cars with less convenience
d. discouragement of heavier cars
e. delivery problems with fossil fuels
f. the impossibility of making a substantial contribution (higher than 50 per cent reduction) to CO_2 reduction.

In Chapter 6 it became clear that carless households can face marginalisation as the trend towards car dependence continues. It also became clear that poorer households may be faced with higher transport costs. Both groups may face problems towards 2030 with accessing important services and facilities, which will be located in places beyond their travel horizons.

The problems mentioned will not arise if:

a. there is huge growth in the energy efficiency of cars, towards for example 40 kilometres per litre;
b. more than 80 per cent of the 2030 car fleet becomes non-fossil fuels related;
c. compensation is made for risk groups on mobility (part of the carless households, and poorer households);
d. physical mobility is replaced by virtual mobility on a great scale.

If these four developments do not happen, or – more realistically – do take place but not on a substantial scale, Western Europe will be faced with the following paradoxical situation 'the car has become indispensable, but driving is becoming more problematic, from different perspectives'. This is a problem. Cars are not just consumption articles but are a carrier in the functioning of modern societies.

Let us do a small thought experiment. If, through some mystery event, all the cars in a country could not be driven for two weeks many things would go wrong; public transport would not be able to accommodate all the commuters, people would stay home out of sheer necessity, meetings would be cancelled, a lot of volunteer help will be impossible, highway locations could not be reached and fewer leisure activities could take place. On the other hand; children could play in the streets, people could get to know their neighbours while at home, public transport would make a lot of money and some stress would disappear. There would also be some pragmatic accommodations; the bicycle would be useful, courtesy calls would increase, clever organisational ideas would be delivered!

In just two weeks we would be able to see how essential cars are to the daily rhythms of modern societies. The car is not only the great connector, it is also 'the linchpin of a whole system of inter-linkages between central sites of production and consumption in our world' (Soron, 2009, 184). And Soron continues 'by any standard, automobiles are absolutely integral to contemporary capitalist society, facilitating economic growth and spurring continual expansion' (Soron, 2009, 184). Society is organised in such a way that the wishes of people can only be accommodated on a practical basis via individualistic approaches. And so the situation arises that 'millions of individual drivers pursuing their rational self-interest in using cars for journeys to work, to shop, and to play, create problems of exaggerated energy consumption and environmental degradation at the collective level – the level of society' (Freund and Martin, in Soron, 2008, 193).

Now we have all the elements for what can be defined as *the great question on car mobility* for the future:

1. The car is essential to keep a differentiated society going; the car connects, binds and integrates in a scattered world. The car is indispensable in reaching the level of flexibility considered normal in our modern societies (which is very high). The car organises production and consumption, and the car helps us to be able to keep consuming, an essential demand in our societies. Without a car people and households will miss chances and opportunities. This will be even more so in the future.
2. The car requires individual choices and individual driving behaviour. The sum of all this individual driving behaviour at societal level is scarcity and congestion, is a substantial contribution to exhaustion of fossil fuels, is environmental degradation, and is a substantial contribution to global warming. It looks like the boundaries for cars using fossil fuels are in sight, certainly when seen from a world car growth perspective.

3. There is, as yet, no winning technology for real and substantial new forms of car mobility that can meet CO_2 and other sustainability objectives and has less or no dependence on fossil fuels.

In one sentence 'The car is indispensable but can face an uncertain future.' In the meantime car dependence is growing in all western societies! This means that our societies grow more dependent on an object and a system that will not remain. This can be seen as 'looking for trouble'.

However, this is not yet a generally accepted idea. Many leaders in the debate on the future of car mobility believe that with very much more energy efficiency, with fast introduction of the newest IT-related vehicle technology and with a speedy shift towards electric driving we will be able to stay out of ugly problems with fossil fuels, we will be able to reach objectives on global warming, and we will be able to keep driving in a more modern and more sustainable way. We will call this the optimistic school.[1] In Section 7.2 we will present two scenarios that fit in this school.

There are, on the other hand, authors that believe that we are at the end of car mobility as we know it. The system that has led to frequent car use, and to car dependence has now, in their opinion, so many anomalies that it cannot be expected to continue, even with great technological changes. These authors believe that all the technology combined will not do the trick to make a sustainable system of car mobility on the world scale. We will call this the realistic school, as these authors look at facts and figures instead of expectations and future perspectives. In Section 7.2 we will also present their line of argument.

Another distinction in the ways to cope with the great question can be made. The processes on car mobility from now on can be seen as a great and important transformation process that should in some way be lead. However, it can also be seen as just an important challenge, hoping that all stakeholders will take some responsibility. With this last attitude, chosen for most societal problems, we will see a transformation dynamic that is unpredictable and chaotic. The first attitude leads to the definition of the shift (whether into technology with the optimists, or to a new system of mobility, with the realists) as a major societal project, with interlinked actions and time schedules. We will call this the governance on car mobility, and we will introduce four scenarios for governance in Section 7.3.

7.2   Optimistic and realistic scenarios

In the *optimistic* scenarios on car mobility we have to do better. Better with regard to the efficiency of cars and in meeting targets for fuel efficiency. Better towards the reduction of CO_2 and global warming, attaining global objectives. We should

---

1   The optimistic school does not, in chief, consider the problems mentioned in Chapter 6.

also make the shift from fossil fuel vehicles to other powertrain technology, such as electricity, all in two decades.

In 2009 Shell introduced the *Shell PKW – Szenarien bis 2030* for Germany. These scenarios were based on the Shell Energy scenarios, introduced in Chapter 5. Related to Scramble is the trend scenario, *Automobile Anpassung*. The core of this scenario is continuous renewal to the upcoming insights. No great innovations are foreseen. Related to Blueprints is the alternative scenario, *Auto Mobilitat im Wandel*. The change here is a jump towards cleaner energy and the introduction of a policy spectrum of policy measures, resulting in more multimodality in passenger transport. International cooperation is foreseen. As Germany loses population in the years to 2030 (from 82.2 million to 78.5 million) the increase in demand for mobility will not be very high but there will still be an increase in car ownership, while the number of kilometres travelled will remain stable. The trend scenario will not reach the objectives on CO_2 in 2030 (minus 40 per cent from 1995) it arrives at 24 per cent less CO2 (1995–2030). The alternative scenario comes nearer (35 per cent less 1995–2030, Shell, 2009, 41), the barrier being the slow change of the car fleet. As Shell state '*Bei einer statistischen Pkw-Lebenserwartung von 13–15 Jahren kann Auto-Mobilitat nicht innerhalb weniger Jahre revolutioniert werden*' (Shell, 2009, 47) [With the life expectancy of cars of between 13–15 years, little can be revolutionised within a few years]. Bringing in new technology in as quickly as possible, and being as realistic as possible, thus accepting that fossil fuel technology will remain for a long time, is Shell's message.[2]

In the UK, the Office of Science and Technology commissioned, as part of its Foresight Programme, a project on Intelligent Infrastructure Systems. In this project four scenarios towards 2055 were developed (*Intelligent Infrastructure Futures, Scenarios towards 2055*, 2010). Two axes of uncertainty were central in the design of the four scenarios. The first was whether we will develop transport systems with low environmental impact, and the second was whether people will accept intelligent infrastructure (elements of driving being taken-over). Basically there are two success and two failing scenarios. The two failing scenarios are Tribal Trading (which describes a world that has gone through a sharp and savage energy shock with long distance travel being a luxury)[3] and Good Intention (wherein the market failed completely, and government has taken over to reduce carbon emissions, with a 'big brother is watching you' attitude). More interesting are the two success scenarios. In the optimistic school there is *Perpetual Motion,* which looks comparable with the alternative Shell Germany scenario. It describes a society driven by constant information, consumption and competition. Demand for travel

---

2   Although this sounds wise; it is important to take note of the fact that in the Shell scenario, even with a reduction of 35 per cent in 2030, it will be impossible to reach 60–70 per cent CO_2 reduction levels in 2050. And there is no attention to the themes introduced in Chapter 6.

3   This scenario contains elements of Blueprints and is comparable with the scenario of the daughter of van der Linde (see 5.4).

remains strong, new, cleaner fuel technologies are increasingly popular. Road use is causing less damage. Urry calls this scenario 'essentially a version of what has been termed "business as usual" or "hypermobility"' (Urry, in *Foresight Scenarios*, 2010, 13). It is, however, unclear how much CO_2 can be reduced in this scenario. There is also no attention for to carless households or to fossil fuel aspects.

Taking into account also the French scenarios *Chronos, Hestia and Pegasus* (as described in Section 5.10) we are able to identify that the most far reaching Western European scenarios in the optimistic school are likely to reach a 50 per cent reduction of CO_2 emission from transport in 2050 (related to 1995/2000), while not taking into account the effects on carless and poorer households, and with an rather unclear treatment of the developments on fossil fuels.

This is where the *realistic* school comes in. They take the 60–80 per cent reduction objective for 2050 for transport and the fossil fuel aspects very seriously and pay at least some attention to the social aspects. To begin with the German researcher Rammler: In his pamphlet *Reinventing Mobility; 14 Theses on Mobility Policy* (2010) he states 'We are at the end of mobility as we know it. Fossil fuel based travel and transport constitute a historical epoch that is now coming to an end. Moreover, the cultural model combining car ownership and suburban lifestyle is at an impasse.' He sees as the biggest problem for car mobility; 'the finite nature of fossil fuels, that should be the major theme in the contemporary discussion of mobility'. Rammler wants to design a new mobility system, and asks for societal energy. It will not be simple. In his idea the following rationale for frequent car use is at work

> the more the paths of time and space are individualised, and no longer follow simultaneous patterns, the more will the degree of autonomy and flexibility associated with the means of transport become a key criterion in making choices …
> In a society where the structures of time and space, variety in life styles and sense of purpose have been developed for decades on the basis of automotive functionality, this device has become a constituent part of the system. (Rammler, 2009, 19)

In his vision the incremental logic of innovation for a 'subsystem perspective' should be 'countered by the ideal of an ambitious mobility policy that targets society as a whole'.

From a somewhat different framework Moriarty and Honnery arrive, in *Australian Car Travel; An Uncertain Future* (2008), at same conclusion. At first they clarify the challenges, climate change and oil depletion. Then they analyse all the offered solutions – fuel efficiency, use of alternative fuels, and sustainable public transport. And confronting the challenges with the solutions they find gaps. In their opinion it will not be possible to find technical solutions for the two challenges. At best a 2.5 times higher fuel efficiency can be reached, and this result could be offset with higher fuel costs and with lower car occupancy rates. And electricity in car mobility will find its boundaries in the non-availability of enough carbon neutral renewable energy to derive electricity from. From their analysis they end with a far-reaching conclusion 'vehicle travel levels will need to be reduced threefold or even more, depending on population growth' (Moriarty and Honnery, 2008, 11).

Urban Colonies, the last of the four Foresight scenarios, belongs to the realistic school. In this scenario investment in technology is primarily focussed on minimising environmental impact. Good environmental practice is in the heart of mobility policy; sustainable buildings, distributed power generation and new urban planning policies have created compact, dense cities. Transport is permitted only if green and clean; car use is still energy intensive and is restricted. Urry considers this scenario attractive; however, he makes a very interesting point:

> How would this scenario come about? It is difficult to see its emergence as being a linear development from existing patterns or something that governments could simply introduce ... There would be some kind of 'shock' to the system and this would almost certainly be a 'global' shock that provides a 'tipping point', a little akin to the global shock of 9/11, ... a global shock that is understood worldwide as a threat to the pattern of 'business as usual'. (Urry, in Forsight Scenarios, 2010, 14)

In fact, after this short introduction of scenarios both schools, the optimistic and the realistic school both seem to have an optimistic and a realistic side.

The optimists are very optimistic about reaching the necessary objectives, but they will not reach the stated objectives with their approaches. They are in the meantime realistic about the chances of realising what they present as their approach.

With the realists it is exactly the other way around. They are realistic about what really should happen to reach the objectives on fossil fuel dependence reduction and on CO_2 reduction. They are, in the meantime, very optimistic about the chances to getting modern western risk societies to endorse their approach.

The difference in governance attitudes is also interesting. The optimistic school wants basically to continue business as usual on car mobility. The realistic school wants a change in the functioning of society and even hopes for a real disruptive situation. Both schools, and here they jointly differ with most government policies, want a transition programme (technological, or societal) to be defined. Here the contrast is between their wish for a programmatic orientation, and the normal incremental, piece-meal orientation. This is a question of governance attitudes on car mobility. We will go into more detail on this important theme in the next paragraphs.

7.3   Governance perspectives on car mobility: An introduction

In the next paragraphs four governance perspectives will be introduced. Governance perspectives are not scenarios – they present attitudes to looking at the near future of car mobility, car use and car dependence. In each perspective the role of government will change. Scenarios describe what can happen, perspectives describe the way decision makers can steer, lead or govern on a problem.

Here we talk about the future of car mobility. It is clear that this future will be sought in a landscape of uncertainty. Looking at the literature the majority of scientists are convinced that some form of transition towards sustainable mobility will be necessary, with some decrease in the role of the car, in relation to other transport modes, with emissions targets on CO_2, and with less dependence on fossil fuels. Most scientists show a global perspective – in their eyes it will be impossible or very difficult to extend the levels of car use in western societies to the world at large, without crossing sustainability boundaries. Outside academia the focus is, with the optimistic school, on technology. Concrete targets are not their business; they present technical possibilities and hope that the technology will accelerate.

In this reality it seems unclear what exactly should be reached in the relationship between sustainability, energy situation and car mobility. What is the core of the change needed in the car system? With so many uncertainties and so much lack of clarity there is room for much debate. Most authors consider a few aspects of the overall problem important, forget about others and base their advice on a small basis, with elements of ideology intricately linked to their vision. The lack of clarity in objectives to be reached gives rise to uncertainty and while this uncertainty prevails the greater initiatives will not be taken by governments, by enterprises, in academia. Minor studies and pilots are created but for the real important initiatives everybody waits for greater clarity.

Teisman (2005) elaborates on this situation. In his view there are today no clear problems anymore, as there are no leading objectives. Clear problems can be defined as 'such undesirable situations that politicians and society feel the urge to act'. The decrease of this type of problems is problematic for governments that want to act. Governments in our western risk societies are now left with the so called 'wicked problems'. The characteristic of wicked problems is that the differentiation between problem and solution is difficult to make. Returning to car mobility, frequent car use and the future problems:

- Is the core of the problem to create a form of car mobility that remains within the boundaries of sustainability? And at which level do we then look at sustainability? In our countries, in Europe, or worldwide (with all its equity aspects involved)?
- Or is the objective to defend our nice and well working systems of car mobility and the related western lifestyles against challenges coming from energy, with fossil fuel delivery problems and possibly global warming as core problems?
- Or do we feel that the car facilitates in essence a lifestyle that creates big risks for sustained global peace?

There is no consensus whatsoever about the greater objectives. It looks like optimists frame their objective more along the second problem definition, where the realists frame more according to the third problem definition. Both schools pay lip service to the first problem definition, but they seem to differ on the scale to approach sustainability.

There is, however, a common denominator. Both schools and almost all authors seem convinced over the necessity of the following aspects:

a. it is useful to reduce CO_2 emissions from car use
b. it is wise to invest in less dependence of fossil fuels
c. increasing fuel-efficiency is essential
d. it is useful to build lighter cars
e. it is wise to bring more IT into in-car systems (for capacity and for safety)
f. it is useful to replace physical mobility with virtual mobility.

The opinions on time schedules, forms and priorities differ, but this is the 'bottom line'.

From this bottom line we will elaborate four steering perspectives on car mobility. A steering perspective describes the way governments would like to intervene in a problem area. Steering perspectives indicate basic attitudes for framing problems.

For the design of these steering perspectives two axes are introduced. The first is about the scope of the perspective. Is it just about car mobility coming to grips with the challenges of energy, climate and sustainability (the technical scope), or is the scope broader, also taking into account the social and societal questions?[4]

The second is about objectives. Is the steering perspective primarily working towards safeguarding driving, thus keeping all the advantages of car mobility, or is the steering perspective working primarily towards reaching sustainability, or are both objectives at stake.

Combining the two axes creates the four steering perspectives.

Table 7.1   The four steering perspectives

Scope	Objective Safeguarding driving	Objective Sustainability	Name Steering perspective
Technical scope	+	+	Following Wisely
Technical scope	+		Changing in Optimism
Technical and societal scope		+	Transport Transition
Technical and societal scope	+	+	Creative Complexity

---

4   As introduced in Chapter 6.

The first steering perspective is the technical perspective for both objectives. This is the actual steering perspective of the Dutch government and of most Western European governments. Fuel efficiency is seen as important, as is the introduction of electrical vehicles. No choice is made between safeguarding driving and sustainability; the idea is that you can have them both. New developments are being monitored. There is no communication on real transitions to be made. Developments are followed, sometimes with an intervention. This perspective is called *Wise Following*.

The second perspective is the technical perspective with a focus on safeguarding driving. This objective is seen as more important than achieving sustainability. The perspective is based on the lack of urgency regarding global warming – there is enough time for change. And the change is towards more efficiency, more IT in cars, and using fewer fossil fuels. New and better technology should do the job. There is no rush, car use is seen as positive for maintaining the fabric of modern society. Optimism prevails. This perspective is called *Changing in Optimism*.

The third perspective has a broader scope, and the objective of reaching sustainability is more important than safeguarding driving. In this perspective it is expected that problems with the delivery of fossil fuels and/or climate change will become actual over the next decade. There is no time to lose; we need to invest in a shift in our focus on mobility. Driving should be brought within the boundaries of sustainability. Sustainable mobility will mean less car use, and will need a complete transition process. This perspective is called *Transport Transition*.

The fourth and last perspective has the broader scope, and aims at both objectives. The urgency regarding fossil fuels and climate change is understood but the intervention is more open than in the third perspective. The end result is open and flexibly defined. In this perspective a portfolio of possibilities and chances is shown. The 'change energy' in modern western societies is seen as the main driving force. All interventions are aimed at reaching and creating energy for change. New arrangements and new combinations of stakeholders can create this energy. This perspective is called *Creative Complexity*.

In Sections 7.4 to 7.7 these four steering perspectives will be presented, and a broad overview on governance possibilities on car mobility will arise. Section 7.8 will show a synthesis on the basis of the four perspectives, and in Section 7.9, the focus will be on governance capacity in society, as the condition to create the necessary changes.

7.4   Following wisely

This steering perspective is primarily technically oriented, and makes no choice between reaching goals on CO_2 and the need to keep driving. Both objectives are important. The core of the approach is to cope with problems when they arise, for example, congestion is a current problem, climate change is a problem that we should probably deal with, and as delivery of fossil fuels is not yet a problem we are not working on major policies. In this perspective following all developments is an important focus. For each development that can have effects a first, small basis for some form of policy is created. For example, in the Netherlands there is a pilot scheme using electric cars, and minor policy on mobility management in association with employers. There is a variety of minor policy interventions. Because one can learn from the experiences of these minor policies, we can give flexible answers when real problems arise, and a statement originating from this steering perspective. The basic attitude is to move gradually with developments.

Is there a content-oriented strategy in this steering perspective? Car driving and car mobility are not approached in a normative manner in this perspective. The idea is that car driving is useful, fits society, has some disadvantages and some risks, and these we have to minimise or to mitigate. The steering perspective accepts the goals on CO_2 and waits for extra steps to be taken at EU-level. Energy efficiency is a goal on which is worked in a non-partisan way.

The core theme in this steering perspective is accessibility, or in Dutch 'bereikbaarheid', meaning; being able to reach destinations without delays. The real policy focus is still on creating enough capacity to make this happen. Essentially the Dutch mobility policy, and this holds true for many Western European countries, is a capacity policy or better – an infrastructure-oriented policy. A real mobility policy would put questions like 'why this demand for mobility?' and 'are we going to accommodate all mobility?' at its core.

Capacity and infrastructure policies take the demand for mobility for granted and focus on ways to create capacity for this demand, over and over again. Creating capacity is done in clever ways, not only by creating new infrastructure, but also via traffic management and mobility management. The core policy of this steering perspective is the cleverest form of what is called in the literature 'predict and provide' policy (Vigar, 2001, Hickmann and Bannister, 2006, Paterson, 2006). The demand for mobility is seen as a black box. A focus on accessibility of destinations and services for vulnerable groups, as introduced in Chapter 6, is missing in most Western European countries. *Mobility policy with its capacity or infrastructure orientation is essentially policy for the huge middle classes in our Western European societies.*

Still this creates some problems. Accessibility is used as a neutral policy term. But sometimes the question 'accessibility for whom?' needs to be answered. Should investments be focussed on all car drivers, or should we concentrate on areas with accessibility problems, where the most 'value added' in economic terms is created? Leaving these questions somewhat out in the open creates an intuitive basis for mobility policy, but also leads to problems with the design of specific infrastructural projects, in specific areas.

In essence this is the dominant steering perspective in most Western European countries. Capacity and infrastructure are the real issues, and while climate change is not completely forgotten, there is little to no real mobility policy, focusing on questioning demands.[5]

---

5   In the Netherlands a new element could have been a country-wide pricing of car mobility, but this policy has recently stopped, before its introduction.

In such a steering perspective some elements are lacking. The first is a feeling of urgency. There is no message, in a non-technical sense, that car mobility will have to change fundamentally in the near future. The policy system looks rather quiet with a variety of pilots and minor interventions, following the EU on CO_2 and an important programme for creating capacity. The system is however flexible – minor interventions and pilots can be upgraded, when needed. Sustainable mobility is introduced as a goal, but there is no real value-laden and politically-driven programme on this objective.

Also lacking is energy. There is no great enthusiasm for this steering perspective. Stakeholders, academia, the business world do not feel challenged. The policy system is complex, and to a certain degree subtle. There are, seen from the perspective of car drivers, strong advantages to the current system of car mobility; convenience, comfort, flexibility, freedom. And governments earn money via taxes. Changing this system by changing the policies is seen as a threat to society.

One last element missing is clarity. Decision makers supporting this steering perspective give their views on new perspectives such as virtual mobility, in-car IT services in car creating new driving experiences, leading to less congestion, new developments in power trains. But mostly this 'management by speech' falls flat when time schedules for these developments, related to goals to be reached on CO_2, or related to the situation of fossil fuels are to be presented. There is a lot of 'hope value' in this steering perspective!

Why is this ambivalent steering perspective the dominant steering perspective on car mobility in most Western European countries?

The first reason is that the approach can be seen as reasonable and relatively well-balanced. The evident advantages of car mobility are acknowledged and policy makers will not give up these advantages too early for uncertainties.

A second reason is that there is a lot of trust in the resilience of the actual car system. Most decision makers expect that when problems arise – with CO_2, with fossil fuels – these problems will be solved when necessary, through new technologies. Although they are not able to specify what or when, such a feeling is the basis of this steering perspective.

Then there is a lack of imagination. The car, and more widely, the car system has grown so 'embedded' in modern western risk societies that a real change cannot be imagined. Nearly anybody can think without the car. The Norwegian researcher Holden said on this issue 'the transport system is often perceived as an unavoidable and unchangeable structural necessity' (Holden, 2004, 48). He notices households becoming frustrated because they want to be sustainable with regard to mobility, but find no way to achieve this within their lifestyle.

There is also in our modern societies a preference for pragmatism. Policy makers and stakeholders often like to split up the mobility problem into smaller separate problems. Solutions are then sought for these smaller problems. By doing this no urgency arises for a broader transition. This approach fits with the general idea in modern societies that managing the problem is solving the problem!

The final reason is the lack of inspiring vision on what mobility should look like. Many advocates of sustainable mobility present long-standing solutions, such as more public transport and more cycling, solutions that clearly do not attract the majority of the drivers. To quote Kohler (2006) on this issue 'many ideas tend to lack the societal element ideas of how people will live their lives and they also lack ideas for strategies for long term fundamental change.'

All these reasons clarify why decision makers support a variety of minor interventions, and follow the EU on climate policy instead of introducing urgency on a transition of car mobility. *Hoping for technology takes the place of connecting fossil fuels, climate, accessibility problems and equity problems in a well-balanced and inspiring vision on the future of car mobility.*

This steering perspective will probably fail when a real change is needed suddenly, or when a situation arises that was not expected. Then the lack of urgency in this steering perspective becomes a serious backlash, as part of the change energy is difficult to reach. To quote Gorris and Rietveld (2007, 4), 'The transport system is in a lock-in situation, and no individual stakeholder has enough decision power and convincing quality to get us out of this position. Actual structures offer too few possibilities to create solutions. A more fundamental approach is needed to reach a sustainable and resilient mobility system.'

7.5   Changing in optimism

This steering perspective is technical oriented and normative; being able to keep driving is more important than reaching the objectives on CO_2 and fighting global warming. Driving is seen in this steering perspective as essential for preserving the achievements of modern societies. These achievements are valued as very positive by the majority of the populations in western societies and the car is seen as the epitome of a good working system. To quote Webber (1992, 274) 'cars are popular because the auto-highway system is the best ground transportation system yet devised'.

To maintain our society, with all its possibilities and chances, and to develop our society further it is seen as necessary to optimise the conditions for car use, car use being the great connector in our societies. The car is one of the most important basic elements for societal cohesion: Necessary and indispensable. Society has unbundled activities and their location, has spread out over greater distances. Sprawl is seen as positive in this perspective. People want to live in large spaces, not very near to each other, and the build-up area is still a small percentage of the total area in our societies. The disadvantages diminish when looking at the advantages. In *The Sprawling of America: In Defense of the Dynamic City* (2006) Staley states 'Sprawl is not random, irregular or chaotic. On the contrary, land development is constrained by consumer behaviour and production costs ... people want better housing, safer communities and easier access to normal, everyday living such as shopping and recreation' (Staley, 2006, 65).

Car dependence is in this steering perspective considered just a reality, and not as a problem or a potential problem. Organisations working from this perspective just take the time to create smarter cars to mitigate possible disadvantages.

The urgency of climate change is disputed and contested in this perspective. Protagonists are not convinced of global warming, and even when convinced people working in this steering perspective doubt the time frames put forward in conservation and environmental conferences. The urgency to change from fossil fuels is also disputed. People believe in the strength of other exploration locations, the locations for the non-conventional oil, such as the Canadian tar sands. But there is concern about the future of conventional oil – western societies are seen as too dependent on oil states with not to be trusted governments. The possibility of problems of non-delivery of fossil fuels is seen as a serious threat, needing fierce international policies.

The car offers freedom, creates happiness, and makes it possible that households are living where they want to live. 'The car created the opportunity, for the first time in history, for ordinary people to follow their own wishes' would be a sentence fitting in this steering perspective. Balaker clarifies in *Why Mobility Matters* (2006) the range of extra choices the car made possible.[6] He stresses the role of the car in the social fabric in the United States, and points at the situation that many modern lifestyle patterns, especially in the suburbs and in rural areas cannot take place without the car.[7]

To carry this argument a little further. A core element in modern societies is the permanent high level of consumption. We cannot live in modern societies without a whole spectrum of consumption wishes, for which we need a car. *Jackson* creates this vision in a series of articles on sustainable consumption. In his vision consumption has a central role even in the conceptual basis of modern societies. Discussions about consumption patterns lead very quickly to discussions about lifestyles. Changes in consumption patterns challenge vested interests. A quote, from before the credit crisis:

> in particular, under current market conditions, it is almost impossible for businesses to engage seriously with any discussion about reducing levels of consumption. Whereas, in the early days of capitalism, corporate charters emphasized that companies existed to serve society, the rules of the market are now such that they must compete ruthlessly for survival. (Jackson, 2004c)

Consumption has another important role to play in modern societies. Each society has a gap between the real and the ideal, or better stated – between our aspirations and our daily lives. Three strategies are available to bridge this gap;

---

6   Published by the conservative–libertarian think tank Reason Foundation in Los Angeles. This think tank has a relatively broad and interesting transport research programme.

7   It is interesting to see how seldom these statements are made in the official transport research literature. They dominate on the right wing side of the American political spectrum.

naïve optimism, cynicism, and 'displaced meaning'. In the last strategy – the term is from the philosopher McCracken – the ideals are moved to a moment further away in time. Religion fulfils this role, but consumption can do the same, more instantaneously. Consumption, with its ideals and its images, gives us the feeling of being taken somewhere that is broader than normal daily life. Jackson says 'we consume in the pursuit of meaning'.

The followers of this steering perspective want to maintain these circumstances and see three threats to ubiquitous car use. Governments introducing extra regulations, mostly from sustainability perspectives, higher prices for fossil fuels, and congestion, which makes driving more tense, less pleasant and less reliable. The role of government policy is often seen by them as incomprehensible. As it is clear that people value car use, policy starts to restrict the use of cars, with objectives that are less valued by most households. Protagonists of this steering perspective point at the fact that the car, and what it delivers, is more popular than environmental or sustainability issues. The choice for politicians should be clear, in their view.[8]

Three solutions are offered to congestion and for sustainability. The first is energy efficiency: Important for industry and for car drivers, who should not be confronted with higher prices for driving as an end result. The second is stimulating teleworking and working from home (see Balaker, *The Quiet Success; Telecommuting's Impact on Transportation and Beyond*, 2003). The last one is a package on congestion. Fighting congestion is urgent in this steering perspective. Staley introduces in *Mobility First: A new vision for transportation in a globally competitive 21st century* (2009), a combination of road building and pricing mobility, in which the revenues from pricing have to be ploughed back into road investment funds.

Less urgent in this perspective are changes in vehicle technology – much value is attached to heavy, comfortable and safe cars – and realising sustainability. A rather slow transition to sustainable mobility is advised, with the argument that going too fast will upset intricate balances in the social fabric of modern, prosperous societies. And the bias is on technology. Much is expected from new IT-based in-car technologies. One expects that these technologies will be available in time. As we already saw protagonists feel there will be enough time to make the change. Extra argument is that in this steering perspective transport does not need to fulfil objectives in the range of 60–80 per cent CO_2 reduction in 2050. Mobility is crucial for economic development, and transport is derived demand, so the activities that demand transport should get higher targets. Transport should not move faster than technical possible. Restricting mobility is no option. In the basis one hopes to be able to drive as we drive now, in energy more efficient cars, with far more IT services on board.

In this steering perspective some attention is given to the social aspects of car mobility. For households with accessibility problems a car use subsidy is proposed, as mentioned in Chapter 6. And possible future problems with the price of mobility

---

8   In *Mobility Contested* (2005) Dunn calls this even an ethical question and criticises the American 'anti-auto vanguard' as standing outside reality; non-democratic, but powerful.

for the middle class households in suburbs and non-edge cities are foreseen.[9] The Brookings Institution shows in *The Suburbanization of Poverty* (2010) that with the financial crisis poverty is now growing faster in the suburbs than in the cities. Forty-two per cent of the American households live in the suburbs, with 33 per cent of the poorer households. Twenty per cent of the Americans live in the cities, with 26 per cent of the poorer households. In this steering perspective the plea is for lower taxation on driving and on shifting subsidies from other transport modes, like transit, to the most popular mode, the car.

This steering perspective, although American based, is also popular in Western Europe, especially by business men, and at the right wing of our political spectra. In the terms of Motivaction, as introduced in Section 3.1, supporters for this steering perspective will be found on the axis New Conservatives–Modern Bourgeois–Social Climbers–Convenience Oriented. This steering perspective plays however a minor role in policy debates.

7.6   Transport transition

The former steering perspectives are actually functioning in societies. Following Wisely gives insight into the struggle with car use in Western European countries, whereas Change in Optimism describes the American situation.

The two perspectives we will now introduce are different. They are technical and societal oriented, and they are not yet implemented on a larger scale. These steering perspectives find their sources in academia and are discussed in policy circles.

In the perspective Transport Transition the urgency of addressing delivery problems of fossil fuels and the threats of global warming are taken very serious. There is no time to lose. The car is seen as a product with many advantages, but also with a spectrum of disadvantages. The idea that the whole world will take over the driving patterns in OECD countries is seen as completely incompatible with realising a sustainable society at world level. This perspective is normative; reaching sustainability in mobility is seen as more important than being able to keep driving.

Organisations and stakeholders, working from this perspective, are convinced of the non-sustainability of the actual system of car mobility, based on fossil fuels. Important terms are 'path dependence' and 'lock in'. The current system of car mobility has existed in its present magnitude for some three decades and has served its purpose. The development of this system was successful. What then often happens is that solutions for dilemma's and real problems are being framed in conformity with the logics and the approaches of that still-successful system. This is the way 'path dependence' starts. Decision makers do not look without prejudice to the best approaches to the dilemma's and problems, but try to mitigate the dilemma's, or try

---

9   See for example *The War against Suburbia* (2010) where Kotkin criticises the focus on cities in the recent American national policies.

to weaken the problems within the logics of the dominant system. Path dependence can lead to 'lock-in' where the system is in a dead end street.

The protagonists of this steering perspective see the car system already in this dead end street. In *An Interdisciplinary Perspective on Dutch Mobility Governance* a group of young Dutch researchers concluded that the present car system does not have enough self-generating capacity for solving actual and future dilemma's and problems (Avelino et al., 2007a). Stakeholders, but also car drivers, feel powerless, and do not know where to start changing the system. In their eyes the great variety of problems and risks related to the mobility system is a clear signal that crisis is already at stake. However, as a result of the complexity of the system this does not lead to action, but to inertia.

Pel and Teisman state in *Governance of Transitions as selective connectivity* (2009b, 4) 'there is a societal dynamic that surpasses all measures to mitigate auto mobility growth'. Society is formed around and alongside the car, and this is a process that reinforces itself. Car dependence is a path dependence phenomenon. To get away from this path dependence a coherent vision is needed on the intricate interplay of forces creating this car dependence, but this vision is missing, and 'at the same time, this structural problem diagnosis points out clearly that isolated measures are unlikely to counter the self-reinforcing dynamic. There is no panacea; technological efficiency improvements are overhauled by growing volumes, moral appeals lead to cynicism more than drastic lifestyle changes, and spatial densification does not determine people's displacement patterns' (Pel and Teisman, 2009b, 4).

Most protagonists of this perspective use sustainability in a strong ecological sense. They see many problems combined related to car use for the next decade – fossil fuels problems, global warming, disequilibria in society through accessibility problems. A transition to sustainable mobility is in their view a broad societal process, also because the trust in technology is rather low in this perspective.

For this transition they formulate transition management. This can be seen as a directed realisation of a new socio-technical system that can better cope with future risks and challenges than the present system. Transition management is a deliberate process aimed at structuring government activities in such a way that a faster change towards reaching sustainable systems can be reached. Transition management is about giving room to precursors, to early adopters, about creating coalitions between stakeholders that are convinced of the need for change, and about creating transition arenas, when new approaches can be discussed.

Transition management can be seen as a form of 'meta-governance'. Important elements in transition management are (see Loorbach and Rotmans, 2006):

- systems thinking across domains, and on different scales
- use the long term as the basis for shorter term approaches
- backcasting and forecasting

- a focus in adaptive learning
- starting with a bundle of options for change.

Transition management takes time to introduce. The reality is formulated by the German researcher Voss 'Conceptual advance that proposes ways to achieve policy integration, long term orientation, and learning strategies finds it hard to be taken up in practice ... Ideas do not seem to fit in the governance regime which is based on specialisation, shorter range orientation, prediction and control' (Voss, 2003, Voss et al., 2006).

In transition management a structuring is made in the systems that have to go through a transition. Three terms are used:

- *regime*: this is the actual dominant system, with its set of rules, agreements, arrangements and institutions;
- *landscape*: this is the wider area in which the system functions. Changes in the landscape create pressures on the system.
- *niches*: these are protected, or experimental spots within the system.

The actual regimes are the basis for the functioning of a domain. Institutional arrangements and formal rules are made, the accompanying infrastructure is created and maintained, and commitments and vested interests keep the system stable. Because vested systems have so many stabilising elements radical change from within the system is difficult.

Regimes can change from the landscape, or from the niches. Changes from both at the same time are also possible.[10] At first the transition management thinkers stressed change from the niches, but gradually they noticed that transitions do not follow exclusively the pattern of pressure from the niches towards the regime. The attention on processes on the macro level is growing.

Looking at the transitions around passenger mobility it is useful to know what the regime is, what landscape is important, and where the niches can be situated.

The dominant regime is the car, the car system, frequent car use and car dependence, as described in Chapter 4, in the societal stories and the driving forces. With this regime belong wishes, expectations, physical layouts, infrastructures and dominant behaviour. The core element is accessibility for the huge middle classes.

The regime is strong, and has not changed much in recent decades. The problems that the regime raises – energy inefficiency, congestion, loss of playing spaces for children, noise, air pollution, loss of quality of public spaces, loss of cohesion on neighbourhood level – are there also for a long time, and are understood and

---

10 On the different routes for change, see Geels and Kemp (2006).

mitigated, but not really solved.[11] Accessibility and at a minor level CO_2 are taken serious, while accessibility is framed for the middle classes.[12, 13]

At the landscape level it is probably the CO_2 emissions and climate change, and the situation around fossil fuels that will create pressure on the regime. In recent times the effects of the credit crisis could also be mentioned. This does not mean delaying the decision to buy a new car,[14] but the increasing share of the costs of driving within diminishing household budgets. The last element on the landscape level can be the reaction of society to far higher congestion levels, with pricing not being introduced.[15]

At the niches we see all the problems not taken completely seriously by the regime actors returning. Pilots are set up to try to create more environmentally and child friendly school runs, to give children playing spaces back, to renew the quality of public spaces, and help constructions for carless households are developed. This all remains rather marginal.

In the eyes of the regime actors the more important niches connect to acknowledged problems and landscape developments. There are pilots to reduce energy levels, to avoid congestion by spreading starting times from home, on electric cars. Some work is done on mobility management, with employers proposing a mobility budget per employee. In the niches are also located decade-long plans for teleworking, working from home and car sharing. This can be seen more widely – many niche activities are already long-standing niche activities. There have not been introduced into in the system, but are still supported by governments. Pilots exist for everything in mobility policy but no changes are made to the regime.

Special mention should be made for initiatives to create integrated spatial designs. Planning for housing areas, for work zones, and for traffic infrastructure should be combined in integrated concepts, fitting together in time and space. We already saw that these themes are popular in the academic literature. Gorris and Rietveld (2007) show that fine-tuning between spatial planning and mobility policy is rather weak; the effect of two separate domains with different societal frames. In the Netherlands spatial planning is far more government-driven than mobility policy.

----

11   Real results can be seen only in traffic safety. There are far less casualties than three decades ago, despite growing traffic levels. At least, this is the situation in Western Europe.

12   About accessibility see the thesis of Geurs (2006). Geurs notes that accessibility (Dutch; *bereikbaarheid*) is not, or only implicitly operationalised or described.

13   Within the regime in the Netherlands only one structural system change has been prepared. This was pricing mobility. By changing the way to pay for mobility it was expected that congestion would decrease.

14   Delay is disadvantageous from a CO_2 perspective. Newer cars are more energy efficient.

15   However, here the credit crisis leads to far lower congestion figures in the last year, at least in the Netherlands!

The current situation on passenger mobility in Western Europe seems to be:

- a few real challenges at the landscape level;
- many different niche developments, with a great number of them long standing;
- a regime already long in power, not taking all problems seriously.

The situation that almost all regime actors see 'sustainable mobility' as the road to travel is a little disturbing. The term 'sustainable mobility' is not coined well (for a view on the chaos and disturbing situations on the definitions, see Avelino, 2007b). Most regime actors act for the steering perspective Following Wisely, and not for Transport Transition.[16]

Transport Transition is a steering perspective for fundamental changes in the system of car mobility. On paper it looks impressive. However, little has been achieved and some basic starting points are under discussion. Loorbach, one of the protagonists of this perspective writes in *Transition Management for Sustainable Development* 'theories of governance developed over the last 15 years are highly descriptive and analytical and rarely offer a prescriptive basis for governance' (Loorbach, 2010, 162). Shove and Walker notice in *Caution! Transition Ahead* (2009) that transition management has strong roots in the tradition of systems thinking and uses the optimistic assumption that a directed intervention on reaching sustainability objectives can be possible and effective.[17] There is only minor orientation to the fundamental ambivalence in reaching sustainability. For example; what is social sustainability exactly, and how is a balance between ecological, social and economic objectives to be reached? They noticed that the social aspects of sustainability are very seldom studied inside and outside transition management and indeed, the basic texts on transition management (see Schot, Geels, Loorbach, Rotmans and Kemp) are more technically and less societally oriented.

What would a transition to sustainability look like? Scenarios can be found in *Sociotechnische Scenarios als hulpmiddel voor Transitiebeleid. Een illustratie voor het domein van personenmobiliteit* by Elzen, Geels and Hofman (2003). Their first scenario shows a relative easy first decade. In the second decade (between 2012 and 2020) societal problems are growing, fuel efficiency is needed, cities get their own transportation systems, making use of electric urban vehicles and the hybrid starts its development. After 2020 the normal car is too expensive to drive. Different cars with different fuel systems are developed and introduced. Petrol stations choose gas instead of petrol, and between 2035 and 2050 automatic

---

16 However, they mostly support research programmes aimed at transitions. This can be seen as communication oriented behaviour, but fits also very well in the core of Following Wisely, with introducing a great variety of options!

17 On this issue Pel and Teisman (2009ba, 2) 'in this transition approach it is assumed that transitions can be managed'.

vehicle guidance is introduced. Three types of cars remain: electric, gas and a small urban car. Public transport becomes important once again. This rather technical transition route is rather bumpy in the coming decades.

In their second scenario the major cities play a central role. From the urban electric car they develop a transit system, with car sharing in its core. This so-called Citrans – system becomes important between 2020 and 2035 and after 2035 normal cars become too expensive and Citrans takes over the car mobility market. A new fundamental mobility concept takes the lead after 2040.

Note that in both scenarios cities play a more important role than national governments. Their ideas on transport and the high prices for car driving are the driving forces. Also take note of the fact that the transition takes times; more than three decades. In between there are certainly discontinuities to be expected.

7.7   Creative complexity

In this steering perspective the urgency of a transition is acknowledged. The sustainable mobility system should be based on an equilibrium of the two objectives. This steering perspective can also not been seen in reality. The perspective finds its basics in the complexity theory, or in network steering. The difference with transition management is the lack of belief in a central steering; targets and possibilities are products of the chosen working methods.

An important notion in the complexity theory is that the governing system should be connected to the governed system. The complexity theory has five domains; complex adaptive systems,[18] dissipative structures, autopoiesis, chaos theory and theories related to path dependence. Central term are also; the non-linear dynamics of processes (project management with defined steps does not work), the self-organisation of users and co-evolution (mutual interaction between different systems of governance). A good introduction can be found in *Managing Complex Government Systems* by Teisman, Van Buuren and Gerrits (2009).

Steering of networks has many insights in common with complexity theory. Network steering[19] is about not choosing one direction early, but selecting options and keeping these options open. It is about organising creative concurrence consortia with the ambitions to work on the objectives from different frames and it is about creating learning environments where experiences with the different frames can be shared. Behind these practices is the idea that wicked problems cannot be managed in a structural way, and that clever interventions are the highest possible to aim at. Chances must be seized when there is energy for change

---

18   See Nooteboom (2006) and Nooteboom and Teisman (2007).

19   Network steering has three forms: interactive policy making, mostly introduced at local levels; there the coproduction of public and private stakeholders, a difficult form because of differences in financial rules and of difference in risk expectations, and finally, the coproduction of government stakeholders.

(riding the fitness landscape). Complexity is sought in this steering perspective, not reduced.[20]

In *Governance of transitions as selective connectivity*, Pel and Teisman (2009b, 2), show the differences between network steering and transition management. Network steering has no limit, richer networks with greater variety start to work, however, '.the very polycentric commitments do not allow for a coherent long term future vision'. Transition management looks more goal-oriented, but it is questionable how a directive to change course, with organisations and stakeholders open to reframe the rules on mobility, can be created in a field where 'wicked problems' are normal.

The design challenge in Creative Complexity is to connect the need for change in the mobility system with the different energies and different interests of the mutual, connected stakeholders.

Introduction of this steering perspective in the domain of car mobility can develop along two lines. The first is to use the self-organisation of the users of the system of car mobility and the second is to create chances and variety to reach breakthroughs in the complex mobility system. In this second line scarcity is an important starting point. Scarcity in different ways; delivery problems of fossil fuels and boundaries necessary through emission objectives on CO_2 can create scarcity in driving possibilities and scarcity in capacity now already creates congestion on our roads.

Four decades ago a useful theory was introduced especially around scarcity. This theory fits in this steering perspective, self-organisation being an important element. This is 'the tragedy of the commons'.

The commons, the shared space in the village, is used so intensively by all local farmers that at times a desolate sandy area has resulted. Seen from the perspective of car mobility we can identify at least three commons, three spaces where quality and sustainability are harmed by the individual choices of individual car drivers.

The first commons is the road network – car drivers use the same roads at the same time every day and this makes parts of the road network temporarily 'unusable'. The second commons is the public space – car drivers use this space as traffic space, become the dominant users and are substantially responsible for its loss of quality. The last commons is the so called 'emission space' – car drivers are together responsible for a growing share of CO_2 emissions. Three situations where car drivers use and overuse a collective space.

The Commons literature started with a paper in 1968 by Hardin in which he explained that common areas without owners, like fishing areas, biodiversity or agricultural commons can become overexploited because 'benefits accrue to individuals and costs are collectively shared'. Since this much-cited article two developments have taken place. First an expansion of the concept; starting with natural resources and natural endowments the concept is now used

---

20   Network steering is the opposite of New Public Management, where complexity is completely reduced to a few management rules.

broadly, for example, for the Internet (see Hess, *Mapping the Commons*, 2008). Secondly, Hardin's pessimistic undertone has been replaced by the search for methods to maintain the commons (see Ostrom *Coping with the Tragedies of the Commons,* 1999).

What exactly are Commons situations? Open entrance to a service, resource or provision seems necessary. Also needed is shared property, or at least no private property. Commons are mostly government properties that are freely accessible. In principle the road network, public spaces and even the emission space (although this is a virtual space!) can be defined as commons. Car drivers can overuse these commons, can create a lack in quality, or can create too high a level of emissions, thus causing environmental damage.

It is questionable whether car drivers, and more widely, the users of the commons, are aware of the damage they create collectively, as the result of their individual choices. In *Hardin Revisited; A critical look at the perception and the logic of the Commons*, Burke states 'environmental problems have become increasingly obscure to individual and societal perception' (Burke, 2001, 464). On the road network car drivers can immediately see what they bring about. They can also see that as a result of their driving and parking public spaces will become less safe and separated, and thus less usable for non-car drivers. The contribution of car drivers on global warming is more difficult to spot.

In the last four decades authors have searched for methods with which the commons could be used to exactly the point where neglect and decline come in. Creating access rights and introducing feedback mechanism, by which users have to pay for the damage they cause seems to be essential. But not enough, as Berry, cited in *The Growth of the Commons Paradigm* of *Bollier* (2006) clarifies; 'We know enough of our own history to be aware that people exploit what they have merely concluded to be of value, but they defend what they love'.

Ostrom carried out research on governance arrangements for the commons. In *Types of Goods and Property rights (*2003) she distinguishes seven conditions for a proper management system for commons:

- reliable and easily available information for all participants on the state of the art of the commons and about the expected benefits and costs;
- a shared view among the participants on costs and benefits of maintaining the status quo, in relation to norms and rules that could reasonably be introduced;
- shared norms among the participants ion what can be considered reasonable, and mutual trust;
- a relative stable group of participants;
- participants living in the area and willing to invest;
- participants are willing to accept pragmatic rules;
- participants can develop relatively soft and careful sanctions.

Car drivers on road networks can meet only half of her conditions. Reliable information should be possible, a shared view is more difficult. An idea of what is reasonable seems to exist, and pragmatic rules are probably accepted. The greater problems arise with the condition of the relative stable group of participants, with living in the area and being willing to invest, and by developing easy and careful sanctions. When Ostrom formulated her conditions she looked at areas and not at line infrastructure. However; also line infrastructure can also be seen as commons. The completely free access to road networks do make car drivers genuine 'free riders'. In the words of Stopher and Fitzgerald in *Managing Congestion – Are we willing to pay the price?* (2008, 10) 'Simply stated, because roads in most places are free at the point at which they are used, there is little or no financial incentive to car drivers not to overuse them'.

Where car drivers cannot immediately act jointly, governments play a role as a starting point and as network managers, also responsible for road maintenance. Driving behaviour can be managed, for example by speed regulations, safety rules, and traffic management and in some countries and cities the introduction of toll roads has been considered. It is interesting that governments are not choosing to introduce access rights to the highway network. *Since in essence completely free access to the road networks, an access that cannot be regulated by car drivers themselves, is at the core of the congestion on the highway network it looks logical to introduce a system of access rights for car drivers.* This could be done with permits, or with time slots. Finally, one can think of a form of time-tabling for the vulnerable roads; *these* car drivers can at *these* times use *this* road.

In this way we can manage at specific times and using IT services, the road network, and more widely, the mobility infrastructure as a scarce resource. People will have to buy time slots and will be able to trade with these slots, thus creating efficient use of the most vulnerable roads.

From road networks we will now move to public spaces and the emission space. Public spaces are areas. The conditions of Ostrom can work here. The most problematic public spaces, with their traffic safety problems, loss of playgrounds, and loss of quality of life through dense traffic are to be found in urban areas. In these areas a dialogue needs to be started, leading to rules of behaviour for the users of these public spaces. However, such a dialogue needs to be started between the residents, and the car drivers, cutting through their living spaces, is seldom sought.[21] Car drivers are guests in public spaces and should behave like guests!

More generally it can be stated that norms and rules for the use of road networks and public spaces do not exist. What is, in essence, responsible use of our road networks? And what is responsible use of our public spaces by car drivers? How should car drivers behave, and who should have priority of access? In most

---

21 Here the Dutch philosopher Van Oenen (2007) offers a conceptual background. Core element for not choosing for dialogue is 'interpassivity'. Modern citizens sometimes feel too tired of all flexibility that is asked of them that they will not invest in connecting to other people, although they consider that the best approach!

Western European countries we have norms and standards on traffic safety and in the sphere of the police (extreme driving behaviour, drunk drivers) but for the rest the road network and, to a lesser degree, public space are 'free for all' locations. There are just no mobility ethics!

As we have seen passenger transport over the road network is the source of 12 per cent of CO_2 emissions leading to global warming. For car drivers this remains an abstraction. There is no connection between the personal choices of car users and insight into their contribution to global warming. A personal emissions budget per car driver could be helpful.

With this Commons approach it becomes clear that car drivers in the now existing degree of car mobility are responsible for temporary (road network and traffic safety in public spaces) or permanent (emissions space, playgrounds in public spaces) harm to the commons. *It is however now difficult for car drivers to suddenly start acting responsibly. First instruments – slots, permits, dialogues, and personal emissions budgets – have to be created. Once these instruments are working car drivers can take personal responsibility by buying, using and trading slots and permits, by being active in dialogues and by lowering their personal emission budgets.* It is interesting to analyse why governments have not chosen to introduce these instruments, but this is beyond the scope of this book.

Finally, there probably is a form of self-organization in which car drivers do not even have to act as responsible citizens. Pel and Teisman introduce such an approach in *Mobiliteitsbeleid als klimaatbeleid of watermanagement; zelforganisatie als aangrijpingspunt voor effectieve beleidsmatige interventies* (2009c). This approach is based on four pillars; making clever use of scarce capacity, interventions by employers, introducing mobility budgets for their employees, a crucial role for 'on trip' travel information, and the use of the so-called Shared Space principle.[22] These four pillars combined will stimulate the creativity of individual road users to find collective workable solutions.

It is clear that his last steering perspective is rather new, and not yet empirically tested. The appeal is the ploughing back of the greater part of the responsibility for the functioning of car mobility to the car drivers themselves. They can choose, but some choices come at a cost.

7.8   Towards a synthesis

The four different steering perspectives define four different challenges. In Following Wisely, the challenge is to create a sober and efficient policy for car mobility in the future. The challenge in Change in Optimism is to keep driving in a time where lots of problems related to car use are taken serious. The challenge in

---

22   Shared Space stands for a form of organising physical space in which the different traffic streams are not directed to their own infrastructure, but are intermingled. It is the need for every traffic participant to be careful (see *Shared Space; Room for Everyone*, 2006).

Transport Transition is to reach sustainability in mobility as fast as possible. And the challenge in Creative Complexity is to create a clever and innovative policy that leads to sustainability without forgetting the advantages of car use. Different groups will be champions of the four perspectives. To put them in line: pragmatic decision-makers, car lovers, environmentalists and innovators.

A pragmatic basic policy

Is a synthesis possible? Or are the four positions so paradigmatic different that we will remain in the current situation – no sense of direction on the future of car mobility. A variety of frames is available, within each frame a policy is available, but there seems to be little 'common ground'. The actual common ground was presented in Section 7.3. Can we reach more common ground? The answer can be somewhat positive. Each steering perspective contains a few policy elements that can be accepted by the other steering perspectives without too many problems:

a. The recognition that there is a chance on delivery problems of fossil fuels, and that it is wise to anticipate this risk in policy (from the steering perspective Transport Transition).
b. A greater attention on the technicalities of a fast shift of the car fleet towards cars using alternative power (from the steering perspective Transport Transition).
c. A far better policy towards stimulating working at home and teleworking (from the steering perspective Changing in Optimism).
d. A focus on possibilities for car users to contribute directly to an effective policy on car mobility for the future (from the steering perspective Creative Complexity).

These four elements could be introduced in the steering perspective Following Wisely, thus giving that perspective a more active orientation. When in this steering perspective adjacent to the four already mentioned elements two extra elements can be introduced we will reach a reasonable broad basic policy on car mobility for the future. These two extra elements are:

• The demographic perspective; many Western European countries are moving towards a decline in population, possibly leading to less car kilometres around 2030 or 2040.
• The situation of the non-car users; wider research on their situation and perspectives in the future would be necessary.

The so defined *broad basic policy* will certainly not convince the paradigmatic proponents of the three more active perspectives. The proponents of Changing in Optimism will state that too much bias is created towards the objectives of sustainability. The proponents of Transport Transition will probably consider the approach far too pragmatic; the real shift is forgotten and the proponents of Creative Complexity will miss the real innovation in this broad basic policy.

This broad basic policy is not more and not less than simply a broader pragmatic common ground, a starting point. It is a policy that will probably not inspire, but is a policy that has a broadness in orientation that fits with the situation that there is no majority yet to be found for a more strict, a more directive policy.[23] This probably is the best to be achieved for now. The proponents of each of the three other steering perspectives can continue trying to convince decision makers on the advantages of only their perspective, but cooperate at the meantime on this broad consensus policy, with now (and this is a difference compared to the steering perspective Following Wisely) an orientation on anticipation.

Towards greater governance capacity

A broader basic policy could probably also be supportive in creating governance capacity. At the moment in Western Europe, governance capacity on mobility is low. 'Governance capacity' is a term used by Innes and Boher (2003, 2010) and by Healey (2007). It defines the capacity of all stakeholders to create joint solutions to societal challenges. This means always reconciling conflicting ambitions and interests. To mobilise institutions and organisation to work towards common defined goals and targets, and to get decisions out of the debating rooms. Easier said than done – this is about the creation of capacity to act jointly!

This governance capacity is now high for some domains and low for others.[24] In domains with a low governance capacity lots of reports are written, lots of research programs are worked out, lots of debates are held, but the end result is just a stand-still, with the same discussions being held over and over again. Regarding car mobility we have to agree with the young Dutch researchers that *the car system has currently a self-generating capacity that is too small for solving actual and future dilemma's and problems.* We will elaborate a little further on this theme from the Dutch context, but it looks like a situation that exists throughout Western Europe.

We defined (Jeekel, 2011, 276) 22 relevant stakeholders on car mobility. These stakeholders can be divided in three groups:

- *the commercial stakeholders*: car dealers, garage owners, the car industry, car insurance companies, oil companies, petrol station managers, driving schools, lease companies, service providers and the providers of travel information;
- *the government parties*: highway or road agencies, juridical services, the enforcing institutions, policy makers and politicians, financial institutions,

---

23   Whether this is towards further accommodating car use, or to stricter sustainability.

24   For example, the Dutch water sector has a high governance capacity.

tax organisations, incident and emergency institutions, municipalities and the regional governments;
• *the societal stakeholders*: employers, road users organisations, environmental organisations, academia.

There are only few systematic connections between these stakeholders. These stakeholders, together responsible for what can be called the system of car mobility, have never been pressed to design together a robust, resilient and future oriented system of car mobility, reaching sustainability criteria and fitting in a broader system of mobility for modern western societies.

Each stakeholder could optimise its own niche. Most commercial stakeholders need many car drivers, driving many kilometres and they do not have institutions that can implement a policy within their group that is, perhaps, not the absolute best for their members but is seen from the common good as the most sustainable approach for the years ahead. Most of them do not accept the possibility that their members would follow these sorts of regulation. Most commercial stakeholders' branch organisations are rather weak.

The government parties differ greatly in their objectives. For the Ministry of Finance heavy vehicles are better than more efficient and smaller vehicles; more tax revenues are the result of bigger vehicles. Policy makers have other objectives. There often is disagreement on traffic issues among the different government layers. The enforcing institutions, the road agencies and the incident and emergency institutions can only follow the rules actually in place.

Finally, the societal stakeholders illustrate the different interests involved in the development of car mobility.

At the moment there are no definite guiding principles for the direction car mobility should take among these stakeholders. Each stakeholder follows its own policy. In recent years, however, there has been some activity. Three 'centres of joint activity' can be found:

• There is a centre around traffic safety, with the enforcers, incident and emergency institutions, car insurance companies, road agencies and driving schools involved.
• There is a, somewhat weaker, centre around congestion, with employers, car users organisations, service providers, suppliers of travel information, road agencies and policy makers involved.
• And there are some initiatives on sustainability, with the car industry, car dealers, lease companies, academia and environmental organisations involved.

Oil companies, garage owners, owners of petrol stations and the financial branches of government do not seem to be involved at all.

Towards a comprehensive governance perspective on mobility; a systemic approach

What is missing is an investment in a joint design for a robust and resilient car system for the near future. It is clear that current governance models will not work, because they ask for too much predictability. Most governance models will split the connected problems into small sub problems, thus not helping to define transitions. In any design all connections between stakeholders should be analysed, and choices will have to be prepared. A system has to grow which has permanent interaction between stakeholders, acknowledging the interests of all stakeholders. Stakeholders should be pressed to come up with solutions, and slowly but convincingly out of the spectrum of possible solutions an effective approach may arise.

For this work a systemic approach can be useful. Related to mobility three different systems can be analysed:

- The *internal system* of mobility; out of driving forces and societal stories (Chapter 4) there is a demand for car journeys. These journeys can be made, or not made, because staying home is wiser (e.g. for financial reasons, virtual mobility) or because there are ethical reasons for not making the trip. These journeys can also be made by other transport modes, when these transport modes can be seen as reasonable alternatives. Optimalisation in this internal system of mobility are possible; ethics, pricing, other transport modes, virtual mobility, creating more capacity.
- An *external system*, influencing and being influenced by mobility. CO_2 problems, other emissions, scarcity of fossil fuel and visions on public spaces can create pressures from outside on the internal system. To cope with this external system most OECD countries hope for technical solutions. When these solutions are not in place on time, the objectives in the external system have to be changed (lower emission reduction policy), or less mobility is needed.
- The *movement into the driving forces and the societal stories*. These are the source of our mobility demands. We could look at the 'storage life' of these driving forces and societal stories. They can also change in time.

It is interesting to see that most work on mobility has concentrated on the internal system. The external system only recently came into view. Very little work has been done on searching for options to change driving forces and societal stories related to mobility and car mobility. There are biases in activity. In the internal system the focus has been on capacity, and on looking at alternative transport modes. Less has been looked at IT in relation to mobility, on pricing, and no discussion has started about ethics in mobility. In the external system CO_2 and air pollution are seen as problems far more than fossil fuel delivery or the quality of public spaces.

We can now define the contour of a comprehensive governance perspective on car mobility. Adjacent to the before mentioned pragmatic broad basic policy approach three elements need extra attention:

a. searching for options to change driving forces and societal stories[25]
b. ethical questions around mobility; what is well balanced mobility?
c. the approach of time perspectives; time and mobility needs elaboration.

There is probably a relation between the lagging behind of these three elements and the difficulties in defining objectives on car mobility in the future. The debate is basically not on the right level.

Although from a pragmatic viewpoint we should now be ready to accept the boundaries set by the pragmatic approach it is clear that the long term involvement of all stakeholders will be supported by clarifications on this deeper level. The best approach is here to work in parallel:

• on the pragmatic steering perspective (broad basic policy) as introduced earlier;
• on clarifying the objectives, by changing the discussion between proponents of the three steering perspectives (Change in Optimism, Transport Transition, Creative Complexity) into a search for common objectives.

Finally, on these common objectives a plea has to be made. It will not be useful, for society at large, to continue discussing whether the objective should be to reach sustainable mobility or to be able to keep driving. It would however help to define as a common objective the creation of 'robustness and resilience in the mobility system to be able to meet future challenges'.

The system of mobility and the system of car mobility should become less vulnerable. Huge dependence on a single transport mode makes a system vulnerable. Diversification is wise when looking at the future and it is probably wise to stop the growth in car dependence.

This does not necessarily mean that there should be no growth in car kilometres. The only new element is that more car kilometres can in the future also be made with other transport modes. The defined common objective is useful from a sustainability viewpoint, and does not immediately ask for reducing car mobility. Here we can probably define the last common ground.

---

25 In environmental policy this is called 'source policy'. Return to the source where demand began.

Chapter 8

The Social and Cultural Aspects of Mobility: Towards a Research Agenda

In this book we have discussed mobility in western risk societies, with a focus on car mobility. Mobility can be studied from a great number of aspects and, for most, many results and many research programmes are available. There is a lot of research on vehicle technology, on the economic aspects of transport and mobility, on energy issues related to mobility and on the spatial consequences of mobility, to name but a few. The social and cultural aspects related to mobility were not paid much attention in recent decades and the same holds true for aspects of governance and politics around mobility.

This book focuses on these somewhat 'forgotten' aspects. In this chapter we would like to introduce a short research programme on these aspects. This programme is not comprehensive but will focus on ten themes that we consider of primary importance in understanding mobility and especially car mobility from a social science perspective.

The ten themes are clustered in three particular domains; cultural, social, and governance. For each of the ten themes we will present an introduction, the state of the art of the knowledge and the most relevant research questions.

8.1 The cultural domain

In this domain I identified four themes.

8.1.1 Convenience as an important driving force for car mobility

Introduction
Convenience can be seen as one of the most important motives for frequent car use. At the same time there is hardly any literature on convenience. Convenience has become normal in our western risk societies. We have grown accustomed to convenience and we can afford convenience. We no longer need to make sacrifices to reach locations; we can travel with ease.

It is clear that convenience as the standard is a characteristic of the richer states in the world. In all countries some households have convenience, but only in the OECD states (and there not even in all states) has convenience been democratised, becoming the standard for most households.[1] It is also interesting to note the growth in convenience standards. Fifty years ago the bicycle was seen as convenient, now the standard for convenience is an air-conditioned car with many gadgets.

Expectations and convenience levels are interrelated. Households simply do not expect to be confronted with what they see as hardship. Take for example the situation of escorting your son and three of his friends to a football competition 15 kilometres away. If one of the fathers decided to go with the four boys by bicycle that probably would not be understood by the boys and by the other parents. Parents who deliver their children to school wet, a usual occurrence 30 years ago, because they walk or cycle with them even in rainy weather, are likely to receive a lecture from the school director or remarks from other parents; this is just not normal!

Convenience is a private good. You have to purchase it. Poorer households cannot afford the spectrum of convenience of richer households. Convenience is not socially neutral.

Current situation on knowledge
Convenience does not seem to be a theme for scientific analysis. A few convenience studies have been published by a research team coordinated by Shove (2002a; also Southerton, 2001) and within consumption studies some attention is given to this theme (Jackson, 2004b, 2006). This situation can be noted as rather strange; working towards convenience and comfort is a driving force in day-to-day consumption. Convenience and the growth in convenience are seen as obvious.

Research questions   What happened with convenience in the last century? How did convenience look a century ago and how could the huge democratisation of convenience have taken place? Can convenience grow further still? Or is there a saturation point? How does the individual objective of convenience relate to higher level societal goals?

At first instance it looks as though there is a conflict between aiming for more convenience and comfort and reaching sustainability. Or is a trade-off possible, with 'wise convenience'? Why do people want to have so much convenience? Is this a deeply felt wish of all people? Or can we note a necessity element? Convenience can, for example, be necessary in a society that requires so much flexibility of its members; convenience and comfort create 'traffic islands' in a stressed existence.

Regarding these questions I will study convenience first. The next step will be to relate convenience to car use. Why is the car considered convenient and is this the situation in all circumstances?

---

1   In western welfare states for some 15 to 20 per cent of all households, life is not convenient. They cannot live up to the usual standards.

8.1.2   From necessity to choice; contacts with your own body, with other people and with the weather

Introduction

In early times moving your body was a necessity. By bicycle, on the bus, walking, you had to move, in full contact with the weather and in full contact with other people in the public space. This still is the case in many developing countries.

It is, however, no longer the case in modern western risk societies. You get your body in the car, sit still and the car moves you. You do not have to be in contact with the weather, or with other people. The only things you have to watch are the road, other traffic and sometimes you have to use your windscreen wipers.

With a car, physical activity is hardly necessary. There is no obvious need to move your body. Obviously moving your body was not always attractive. You had to travel in bad weather but your body obtained all the exercise it needed. Obesity has now grown to a huge problem in modern western risk societies.

Moving developed from need to choice. Car users can choose to have physical exercise. Moving by choice is for the weekends, the evenings and sometimes early in the morning. A whole industry has grown; from running coaches to fitness centres and a whole culture on the heroism of physical exercise has developed; the trained body, the admiration for marathon runners.

Not only has moving your body become a choice, the same holds true for contact with the weather. People do not like the wind, except when they go sailing; people do not want rain, only when it is necessary for their vegetable garden and people do not want snow, except for skiing. Basically we would like to control the weather and a remote control would be fine!

Also, meeting other people in the public space has become a matter of choice. The public space, always ambivalent in character and experiences, can be avoided by using your car, you just travel through these spaces.

Current situation on knowledge

There is much literature on the driving forces behind obesity (Ewing, 2003). The relationship between weather conditions and car use is very little studied. There is much literature about the changing face of the public space in modern risk societies (to start with Sennett, 1974) but the relation to car use is mostly only implied. The social and cultural consequences of moving by choice instead of moving out of necessity are seldom researched; health studies dominate.

Research questions

Will moving out of choice grow stronger? And will we be able to exclude climate influence and weather conditions even further? And will the public space lose the greater part of its users, thus becoming more dangerous in the eyes of the risk-avoiding middle classes?

Is it really a wish of most modern people to avoid physical contact? Or is it a consequence of unintended circumstances? Moving your body is not societally neutral. Poorer households use public spaces more often. Most youngsters in the public space are from poorer households and poorer people cope better with bad weather, as they still have to!

8.1.3   Flexibility, immobility and the car

Introduction
It looks as if we have to become ever more flexible. In modern western risk societies temporary arrangements and contracts tend to dominate over lifelong arrangements. Most people will have to grow accustomed to loss of work, to ever-changing circumstances in their work and private lives. The welfare state is vanishing and the motto now seems to be; develop enough social and cultural capital on your own, and take care of your own power in the labour market. And be flexible!

Flexibility is also important in leisure. People are expected to be always reachable and to be always able to switch activity or attention, when colleagues or bosses demand. This societal demand for flexibility is translated into the need to combine activities within set timeframes. To be able to combine and also to make fast switches in your daily programme certain instruments are needed. The car is the big helper. Flexibility and cars are connected. The car is with you and can be used for reaching activities, locations and people.

Cars, mobile phones and laptops create together the possibility of reaching everybody and everything. We can no longer retreat or retire, can no longer be absent. We are being asked to be present, all the time. People with fast and flexible lives cannot live without cars, it seems.[2]

Current situation on knowledge
This theme is being looked at by social scientists. The relation between flexibility, use of time and mobility is an important research theme in the Mobilities school. Flexibility, the need to be available everywhere, and the resulting stress and hurriedness are a theme for Sennett, Bauman, and in studies of the Dutch Social Research Institute (SCP, 2003, 2004, 2006a, 2010). There are also case studies about the growth towards immobility of elder households (Scheiner, 2006; Davey, 2004). What seems to be missing is a conceptual debate about the use of time in our risk societies.

Research questions
The fluidity of always being present is not evenly spread in modern societies. Many people live slow lives, and feel no need to combine many activities. They still have their permanent work location, and have each day the same commute. The gap between the slower lives and the fast and flexible lives seems to be growing. I want to analyse which persons, which households, lead these fast lives

---

2   This intensive use of clock time, with being available for nearly everything all the time, creates stress and hurriedness, but seems accepted by the majority of western households.

and I want to understand the consequences of hypermobility for the functioning of households, urban regions, but also for the experience of the car. Will cars become second homes for their fast users, with living and working functions?

Persons and households without cars often have problems with living fast lives.[3] There are degrees in immobility in households without cars. How does this immobility function, what are the driving forces, and the consequences? Why are so many elderly households anxious about losing their driving abilities? We should like to learn more about the route from hypermobility to immobility and its consequences for lifestyles and welfare.

8.1.4   Ethics and mobility

Introduction
In many societal sectors (like health care or the food sector) ethical questions are acknowledged. It is rather remarkable that ethical questions are almost absent in the mobility sector. Hardly anybody addresses the question of 'well-balanced' mobility, seen from a sustainability perspective or from the perspective of maintaining and renewing the social fabric of society.

Mobility crosses all sorts of boundaries and environmental boundaries are well known. Many publications can be found, however these are mostly pleas for greater sustainability rather than studies on ethical issues. Households today make journeys that were not considered in the past. We will travel 120 kilometres in the evening back and forth to a party with friends. Making kilometres by car was, in the past, in the realm of scarcity and is now in the realm of abundance. We just take the car, without much thought. We have the money, the car is already paid for, except for fuel and we drive.

Current situation on knowledge
Until recently there was hardly any research on the ethical aspects of mobility but in 2008 Bergmann and Sager published a reader entitled *The Ethics of Mobility* and recently Van Wee published a book on Transport and Ethics (2012). On the right wing of the political spectrum some ethical questions are raised about the legality of restrictions on the right to mobility (Dunn, 1998, 2005; Lomansky, 1995).

Research questions
Is there something like well-balanced mobility? This concept exists for driving behaviour. But for mobility in general? Badly balanced mobility is mobility that costs society and other people a lot and costs more than it yields for you. Most costs are on safety and on non-sustainability.

In our modern societies we seem to think that everything that can be done should be done. Where did this line of behaviour start? And how have expectations

---

3   Probably the students and the richer cosmopolitans (see Section 3.1) are the exceptions here.

developed on what a good life is supposed to look like? How might car mobility change if households had to pay for each journey the price of that journey, presented with its full societal costs? There seems to be a relationship between not knowing what driving costs, not knowing the disadvantages of trips and the lack of ethical debates around mobility. I would like to identify these relations more clearly.

How problematic is it not to be able to participate in an activity because you are unable to reach the location of that activity, while all other participants travel easily by car? How do carless households feel about these issues? Is it just a fact of life, or should a well-balanced society offer some support in these circumstances? What is the current situation on riding along with other people, on asking for lifts? Are there forms of solidarity between car owners and carless persons? And is solidarity needed, or should carlessness be seen as an individual choice, for which only the individual person is responsible?

Mobility and ethics raises the issues of equity. Are the joys and burdens of car use well spread in modern western risk societies? Or is there individual prosperity and collective poverty around the car? And is the fact that some 7 per cent of the holders of driving licenses have driving anxiety a warning sign, or just their individual problem?

8.2   The social domain

In this domain I also identify four themes

8.2.1   *Time and mobility*

Introduction
Time is very important in our societies. Many people have the feeling of having too little time, others have too much time on their hands. Sometimes time pressure or running out of time is seen as a status symbol. There are time-rich and time-poor people and it is useful to identify the characteristics of both groups.

In our societies many activities have to take place in a limited amount of time. Mostly activities have to be done quickly. In a certain sense in our western risk societies there seems to be a penalty for slowness. People buy, via speed, some extra time and the car is used because of its speed, thus creating some extra time for activities other than making the journey. This time winning impact of cars demands further elaboration.

Adjacent to time, schedules are important. Just walking along, having a social chat is no longer the custom for time-poor persons. Most people spend their days in an appointment society. This is especially true for working people.

It is interesting that schedules are mostly concentrated in a part of the day, mostly between 8 in the morning and 6 in the evening, the part of the day where also most travel is done. In Western Europe the 24-hour society does not really exist.

The feeling of time pressure or hurriedness is strengthened because the opening and closing times of most shops, professional services and child day care are concentrated in a few hours, mostly between 9 and 6 (nine hours). These hours are mostly the same as working hours. Governments have failed to create forms of time policy, thus creating some relaxation for time pressured persons.

Current situation on knowledge
There is much literature on time and societal patterns. Philosophers like Bergson and Bloch developed concepts of different types of time. The Dutch Social Research Institute did a number of important studies (e.g. SCP, 2010). About acceleration we have the work of Rosa (2005) and the story of collective patterns, but individual solutions to these patterns, is an issue for Jarvis (2004, 2005), Southerton (2003), Skinner (2003, 2005), Schwanen (2007, 2008) and French authors. Mostly the relationship with mobility is not very well elaborated.

Research questions
An important question is why we came to appreciate doing everything by appointment so much.

Spreading activities over the whole day seems difficult. We keep to the nine most important hours, and do not create more fluidity. Why does this conservatism, leading to congestion, still exist?

Because there are no policies on time and opening hours, individual households create individual solutions for what, in essence, are collective problems. People start buying time, the car is needed, stress is sometimes a result. The lack of policy on time is a topic in need of research; which argumentations and ideologies can be identified as the driving forces behind the situation that the rapid increase in stress and hurriedness in our modern risk societies is seen by decision makers and journalists as normal and obvious, and not as an important societal problem to be solved?

8.2.2   The relations between education, income and mobility

Introduction
Higher educated people are more mobile than lower educated people. The same holds true for incomes. With higher education and higher incomes the action radius of people increases. Their family lives further away, their range of work choices is broader, just by taking a greater area into consideration.

In essence lower educated people and people with lower incomes live more local lives, with a smaller action radius. They often still live in the place where they were born and their perspective is mostly not very much broader than a small circle around that place. When the level of services in this area – of shops, of work possibilities, of professional services – diminishes, or when friends move away, this creates problems. For lower educated and lower income households the decrease in quality of nearby services is often a far greater problem than for their richer and better educated neighbours. For this last group their widespread friendship network is more important than the neighbourhood they live in.

State of the art on knowledge
These conclusions can be drawn from statistical data, combined with in depth studies. Area wide travel studies are sometimes available (see Chapter 2). Adjacent to these data are studies about motility (Kaufmann, et al., 2004) and the literature on travel horizons (Morris, 2006) and the impacts of losing services in cities and rural regions (Larsen and Gilliland, 2008; Bowden and Moseley, 2006) is relevant.

Research questions
The concept of travel horizons ask for further operationalisation. When lower educated, lower income households search for jobs in only a small area this diminishes their chances but also leads the not well-functioning labour market. We need to know why these households search only in a restricted area.

Higher educated households seem to be less willing to fight for their local shops. The lower educated households have to take up the fight but often fail to use all channels open to the better educated. I would like to analyse whether this split in modern societies, when cuts in service levels are at stake, can actually be seen.

8.2.3   Accessibility and social exclusion

Introduction
Accessibility is a core concept in mobility policy. Accessibility is mostly operationalised in an economic framework, via travel time, or loss of travel times, related to before defined standard journey lengths. Accessibility is seldom operationalised as the capacity or inability of people to reach the whole spectrum of essential services (from health care to green groceries!) at relative low costs.

Social exclusion through transport problems does exist, as we noted in Chapter 6. Without a car it can be difficult to reach locations early or very late in the day. With a low income taking taxis can be expensive. The location of hospitals, homes for the elderly, or inexpensive shops is of great importance for lower income households and for disabled people. Peripheral locations make travelling within their budget and physical constraints rather difficult.

Current situation on knowledge
The amount of literature on this theme differs. In Anglo-Saxon countries, with their mostly less well defined welfare states, much literature is available. Far less literature on this theme can be found in the Western European welfare states such as Germany, Netherlands and the Scandinavian countries. The French-speaking world has an intermediate position. Most welfare states have the idea that these problems of social exclusion are minor. However, in most welfare states neutral research on the magnitude of possible problems is missing.

Research questions

A first question is how households without cars make car-dependent journeys. Do they make these journeys, with much luggage, in bad weather, on Sunday mornings, or at night at all, or is postponement, or even cancellation, the order of their day?

How do households without cars reach highway locations, rural areas? Do they organise the journeys themselves, or do they expect travel on demand, serviced by governments? And how independent can non-car users be when car-dependent trips are at stake?

A second question is why these themes receive little attention in the comprehensive welfare states. What do the arguments for neglecting these themes look like? And do households and persons with transport problems make themselves known, and by which instruments?

A third question is about the operationalisation of accessibility. Why is the focus on congestion, and not also on the possibilities for groups of households to reach, for them, essential services.

8.2.4   Community light, feeling unsafe and escorting

Introduction

In modern Western European societies the relations between inhabitants of the same neighbourhood can mostly be characterised by 'community light'. Modern people do want some contacts in the neighbourhood, but in a light form. Most people want to identify with their neighbours in a way characterised by a certain distance in combination with easy, but not very deep, contacts. Community light has grown from the circumstance that most higher and middle class households find their friends and family at a distance from their homes. They do not share the same neighbourhood experiences. Community light is a practical strategy for accommodating relationships with the non-friends in the neighbourhood.

In essence, in most neighbourhoods, neighbours do not know each other very well. They share few realities. Spontaneously giving help or advice has become difficult. Households no longer expect other neighbourhood households to watch their children, which was common in the past, when family and friends often lived in the same neighbourhood.

This loss of this spontaneous interaction comes at a cost. Risks and practical solutions are not shared. Everybody feels alone with their problems. Each household has to find ways to mitigate the risks of living together alone. This leads to most middle and higher class households trying to eliminate almost all risks, especially where their children are involved. They ask for levels of safety and security from their elected politicians that probably can never be reached in the always ambivalent and risk-searching modern western societies. Investing in neighbourhood relations is not seen as a useful solution, problems have become individualised or are put in the hands of governments and other institutional bodies.

Small and individual private introspection leads to a situation where much is perceived as risk. People translate the extremities on crime and threat, reinforced by the media, immediately to their own uneventful lives, and become anxious and vulnerable. Just letting your children play outside, for centuries the normal practice, seems to have become risky business in the eyes of many middle class parents! Parents feel that they have to escort their children to all sorts of leisure activities that take place in controlled environments, and this escorting takes place almost exclusively by car over short distances, causing the most damage per kilometre to the environment.

Current situation on knowledge
Much has been published on the loosening of social bonds, or broader of the social fabric, in modern societies. On the consequences for the art of living together far less has been written. There are studies about the level of resilience of modern middle class households. Furedi in particular wrote about the anxieties of modern parents. Literature on the deeper driving forces of the outcry about safety is available. In the literature there are only few relations with mobility mentioned. There is, for example, very little literature on the huge growth in escorting.

Research questions
The interesting question is how the relations between community light, feelings of vulnerability, anxiety and escorting really work. Why does it seem so difficult to create, as parents together, an environment that delivers a feeling of safety? And why do people immediately translate reinforced media stories to their own lives? Where has their resilience, characteristic of our grandparents, living in far riskier circumstances, gone?

Why do many modern households place the blame for not reaching their expectations on safety and security immediately onto the institutions instead of looking to their own behaviour and how does this all affect mobility and especially car mobility?

8.3   The domain of governance

Here two themes will be selected.

8.3.1   *A source policy for car mobility*

Introduction
Source policy is a concept from the environmental policies, which looks at the real driving forces behind environmental degradation. Source policy traces back, not on the symptoms, but on these driving forces. Emissions from factories are not the focus of a source policy, but the arrangements of international trade, finally leading to forms of environmental degradation.

Source policies have seldom been defined in physical mobility.[4] There are probably two reasons. The first is that much mobility policy is in its core infrastructural policy, or infrastructure planning and the second is that demand for mobility is mostly seen as 'derived demand'. The real demand is not for mobility or transport, but to accommodate an activity sector, that needs, inter alia, transport. The source policy was then framed for such an activity sector.

In my view mobility nowadays is certainly not only derived demand, with the real demand coming from activities. There is mobility, just for the fun of it, but, even more important, mobility has become intractably linked with modern lifestyles and modern arrangements of households, and these cannot be considered to be activity sectors.

A source policy for car mobility can be based on the driving forces and the societal stories as presented in Chapter 4. I see six related sources for frequent car use:

- First is the cultural orientation in modern risk societies. There is a huge demand for convenience and comfort, and everything that can be done, should be done. There is no investing in self-discipline whatsoever. A certain boundlessness dominates.
- Second is the wish for community light, with no shared views arising at neighbourhood level. Connected to community light is the call for safety, for protection of households against all risks of normal life.
- Third is the need felt for speed and flexibility, that is being demanded in work relations, and sometimes in leisure, and that leads to difficult time schedules that have to be followed.
- Fourth are the growing possibilities of cars to create identities for their users. You can show who you are by showing your car and relating to its style.
- Fifth is the trend towards highways as locations for work and leisure activities, a trend reinforced by many local governments.
- Last is the power of habit. People follow standard patterns and when the car is integrated in such a pattern it will be used whatever the circumstances.

It is clear that a source policy for car mobility will hit the roots of our modern ways of living. In all activity sector source policies go far beyond the normal day to day practices. For car mobility this will be the same. The urgency for a source policy on car mobility will grow with the uncertainty of delivery of fossil fuels and with problems reaching objectives related to the fight against global warming.

Current situation on knowledge
The six sources mentioned create together the space for a source policy. Little has been published on source policies for car mobility. A number of studies on societal driving forces have been published (see Chapter 4).

---

4 An exception is, perhaps, the second Dutch Structure Scheme on traffic and transport, of 1989.

Research questions
I would like to elaborate on the six already mentioned sources for frequent car use and I would like to define and to design the first contours of a source policy for car mobility.

8.3.2   The governance capacity in the mobility sector

Introduction
Governance capacity is the ability of stakeholders to create solutions for societal questions, and thus to compromise between different interests, to mobilise resources and actors, and to restructure decisions and directions into formal decision making contexts. Governance capacity is about the ability to act collectively. An indication of a low governance capacity is a complete library on reports and government notes that have not lead to practical and visible breakthroughs.

Many authors note that the governance capacity in the mobility sector is rather low. There are many unconnected initiatives, a necessary scaling up is not seen, and mobility is characterised by path dependence and lock-in situations. Nobody sees real progress in mobility policy. The central themes have already been on the agenda for some 15 years; congestion, pricing, sustainable mobility.

Current situation on knowledge
The concept of governance capacity is coined by Innes and Booher and by Healey, and is used in different societal sectors. Literature on governance capacity in the mobility sector is scarce.

Research questions
Why is the governance capacity in the mobility sector so low? Are there too many stakeholders? Or are there too little stakeholders with real power to implement? Is the sector lacking in common shared objectives? Or is there a situation of equilibrium between the different interests, with the result that nothing happens? Do the problem frames of enterprises, interest organisations and governments differ too much?

There is one theme within mobility where governance capacity did create progress; traffic safety. Why did the stakeholder community succeed for this specific objective? And how can governance capacity in general increase in the mobility sector?

8.4   The ten research themes connected

There are connections between the ten themes. Four lines can be noted.

The first line is that in all themes a *better understanding* of the physical mobility in its social and cultural diversity is sought.

The second line is about *equity*. Physical mobility is not socially neutral; richer households are more mobile, and see a greater part of the world as their living space than poorer households and the same holds true for higher and lower educated households. There are completely different travel horizons. This attention to social inequalities is new in the field of physical mobility.

The third line is the attention to *lifestyles and time*. Flexibility and time scarcity are important in understanding a great part of the demand for car mobility.

The fourth and last line is about *anxiety in relation to physical mobility*. Older people are afraid to lose their mobility. Carless households feel vulnerable in their difficulties in reaching services, shops and leisure activities and parents fear the public sphere and choose to escort their children most of the time.

This research programme should lead to greater drama around physical mobility, being one of the defining factors in our western modern risk societies. Only when feelings and arrangements on physical mobility are understood properly will it be possible to design inspiring mobility policies that will replace the boring and repeating mobility policies of today!

Chapter 9
Summary and Conclusions

9.1 Answering the questions; the conclusions

In Section 1.3 we formulated five questions to be answered in this book. In this chapter we will answer these questions, using all relevant material from the preceding chapters and the Addendum.

Question 1: What is current situation of car use in Western Europe?

It was more difficult than expected to find the actual data for answering this question for the 13 countries that we identified as the Western European countries: Finland, Sweden, Norway, Denmark, Germany, Austria, Switzerland, Belgium, France, Belgium, the Netherlands, the United Kingdom and Ireland. On a fairly general level the transport statistics from Eurostat and the national statistics of these countries could be used. However, to probe further, the basis has to be found in national travel surveys. Not all countries publish national travel surveys and the published surveys differ in definitions, questions asked and data presented. Most surveys are published every five years, making it sometimes necessary to use older data and mostly using data from different years.

Given this situation we chose a pragmatic method. We focused on six countries and one European region with comprehensive national travel surveys; Germany, United Kingdom, France, Switzerland, Sweden and the Netherlands and on the Flanders region of Belgium and we accepted the fact that we had to use data from different years. In the Addendum we published the findings.[1] Sometimes it was possible to go broader and present the perspective of the whole of Western Europe, or even of the whole of Europe.

After a long period of strong mobility growth the growth of mobility in Western Europe has slowed down in the last 15 years. Growth of mobility in Western Europe is now for the greater part in most countries based on growth in population. The growth in the number of kilometres travelled per person, rather high in the period 1960–1995, is now relatively low, or non-existent (United Kingdom and Switzerland). Within the general mobility (globally 36 kilometres per person per day) only the car kilometres are still growing. In general in Western Europe 79 per cent of all passenger kilometres is now travelled by car. The car mobility growth in Western Europe was between 1995 and 2006 on average 15 per cent, ranging

---

1 The data presented here are the average of the seven countries, unless stated otherwise.

from Germany (7 per cent) and Denmark (10 per cent) to Norway (19 per cent) and Finland (24 per cent).

At a broader geographical area we see completely different growth rates in most new EU Members States. Where on average in the EU transport growth was, between 1995 and 2006, around 1.6 per cent, in the selected Western European countries it was less than 1.3 per cent and in the new Member States everywhere higher than 2 per cent. The car growth rates ranged between 1995 and 2006 from Czech Republic (28 per cent) to Lithuania (106 per cent).

Looking at car ownership we see that nearly half of the Western European households have one car, 30 per cent of households have two or more cars and this percentage is growing. On average 21 per cent of households do not own a car. Mostly these are single households, the elderly and very young households. There are far fewer car owners in cities than in the rural areas of Western Europe. The density of driving licenses (81 per cent on average, 87 per cent for men, 74 per cent for women) is a little higher than the car density. Driving a car costs, on average, 15 per cent of net household income in Western Europe. It is interesting to note that while the increase in distance travelled is not growing fast in Western Europe, car ownership is (in the Netherlands for example, 13 per cent versus 33 per cent in the period 1995–2009). This means that the distance travelled per car is diminishing (on average in Western Europe 13,500 kilometres yearly) while the number of cars per households is rising. *Cars are becoming individualised consumption articles.*

What are all these cars used for? Looking at the journeys, the share of car journeys in the total amount of journeys (56.5 per cent on average) is far lower than the share of car kilometres in the total amount of kilometres (79 per cent). This means that longer trips are made by car. Public transport takes 10 per cent of the trips (and 12.5 per cent of the distance), the slow modes take 34.5 per cent (and 7.5 per cent of the distance travelled). Shorter trips in Western Europe are more often made on foot or by cycling, while over 7 kilometres the car starts to dominate.

Cars dominate in work-related mobility (more than 70 per cent of the journeys are made by car). Work travel usually means driving alone. For leisure and shopping the car is used primarily for longer distances. Education is the least car-dependent motive. Because cars have more passengers in leisure traffic this motive is the winner in the total car distance travelled. An important newcomer among the motives is escorting, an indicator that the car is becoming a necessity in the fulfilment of the duties of daily life in modern western European risk societies.

Men have a different car mobility pattern than women. They make fewer but longer journeys. Women make some 20 per cent more journeys between 30 and 50 years of age, but these are shorter. Women more often work part time, and do some 80 per cent of all escorting. Higher incomes travel larger distances, and use cars more often. The same holds even stronger for higher educated persons.

Question 2: Which driving forces, and which big societal stories can be seen as responsible for frequent car use and for car-dependent behaviour?

Car dependence is defined in this book as the situation in which a journey can be impossible or difficult if made by a transport mode other than the car. To present data on car dependence I first looked at the driving forces behind frequent car use. These driving forces were introduced in the Chapters 2 and 3 and clarified in Chapter 4. The concluding ten driving forces do not explain all frequent car use but I consider at least 85 per cent of frequent car use will be explained by these ten forces. A differentiation can be made between more society-based driving forces and more individual driving forces.

To start with the four *society-based* driving forces

Geographical spreading out of activities
Activities, once localised in their vicinities, have in the last four decades moved to other locations. A greater distance has to be travelled to combine activities. The once-compact network of locations for related activities has spread out over a wider area. The car can best bridge the distance, and becomes the great connector between activities. The car made this spreading out of activities possible and is the best qualified to connect different locations.

The creation of highway locations
Many companies grow, are in need of space. In many municipalities this space can be found by restructuring existing industrial areas. This however is a difficult activity, due to the situation that many owners of industrial buildings, warehouses and industrial used land, change rather frequent and are active at great distances from the municipalities.

Municipalities and companies based in them often choose to create new space for industrial activities, offices and storage facilities. Free space is, relatively often, found in the vicinity of, or along, highways. Industrial activities, tertiary activities but also housing and leisure activities move towards the highways. People working and/or living along highways make use of these highways. They are car oriented, not in the least because there are few public transport services along most highways. People living in the vicinity of highways tend to be oriented to all interesting locations along 'their highway' and are not focussed on the city region that they formally belong to.

The desire for flexibility
In our modern societies people want to have active lives; they want to work, to care for their children, help friends and family with health problems, and enjoy leisure activities. Alongside all these activities they have to perform daily tasks and normal housekeeping chores. All these activities have to be undertaken in a chain, and in tight timeframes. Most activities have well-defined starting and ending times and professional and specialist services are only to be obtained during daytime. The

start and end time, opening hours of shops and services, have not changed much in recent decades, meaning that most activities have to be squeezed into the hours between 8 in the morning and 6 in the evening. Coordinating activity schedules is essential, as is managing deadlines. Households have to rely on flexibility and speed. Cars and cycles can offer flexibility, the car can also offer speed, at least in most areas. Public transport is just not flexible enough to be able to provide facilities for chained trips.

Creating possibilities

Our societies have many chances and possibilities, which are located somewhere. You should be able to reach them. For example adolescents in the suburban US, have, once their school bus has left, few possibilities to participate in social and sporting activities without their parents' help. A small action radius on mobility makes getting and rewarding chances difficult. Unemployed people without a car have difficulties reaching work on highway locations. Getting your driving license in the US is liberating and older people fear the moment they have to stop driving. In the Netherlands and in Denmark, this driving force is often overlooked, because cycling can be an alternative for a great number of trips.

We will now turn to the description of the six driving forces from a more individual level.

Convenience and comfort

With growing prosperity modern people have the chance to buy convenience. Growing prosperity also leads to a broader spectrum of consumption articles in households, articles that all have to be used, replaced, repaired, and brought. The car is convenient and comfortable for people and travelling with luggage is not very comfortable when done by other transport modes. The car makes it possible not to be influenced by bad weather conditions on your journeys.

The wish for 'community light'

This driving force can be seen as the 'twin brother' of the geographical spreading out of activities. People are looking less for contacts in the vicinity of their home. Most family and certainly most friends live at greater distances. People like to meet others in their neighbourhood via the principles of 'community light'; they will identify with their neighbours in a way characterised by a certain distance in combination with easy moving contact. There is not much investment in community light, so neighbours feel no need to watch other people's children. Using public spaces in the neighbourhood is, in particular in the middle classes, considered with anxiety. Middle class households like driving children to the houses of friends and to hobbies where they can play in controlled environments, and visiting friends and family now means travelling greater distances than in the past.

The lack of ethical boundary

This driving force can be seen as the 'twin brother' of creating possibilities. As people see many possibilities and chances there seems to be no boundary. 'Everything that can be done should be done', could be a motto in modern western risk societies. More people than ever will probably define their lives as a chain of events, spontaneous or self-created. There are also many expectations; being mobile is such a customary practice, that we are now more or less expected to visit activities at locations 100 km away and drive back the same evening. As we see a discourse on thoughtful food, there is no start to define thoughtful mobility. With mobility, everything is taken for granted.

The car for identity

A modern western society has many abstract rules, standards and laws. These societies look ordered. People fit in the patterns created by these rules. However, especially in leisure time they want to search for their own identity, unhindered by rules and laws. Cars can be part of their identity. The, in its broadest sense, consumption of the car, can compensate for many unfulfilled expectations. Many people cherish their cars, the car is the place to be and the body to travel with.

Fears, feeling vulnerable and, and seeking protection

In our risk societies people are aware of the risks related to their lifestyles. In the Second Modernity, risks are man-made. People try to mitigate risks, and to avoid risks. People are living individualised lives which mean standing alone when facing risks. Cars can protect people and households. Cars are great friends in the individual fight with the dangers of public spaces. Travelling by car makes you feel less vulnerable. Cars can be used as a defence mechanism.

The power of habit

The last of the 10 driving forces is relevant to all ages. Living without habits seems impossible. Habits arrange the complex worlds in a structured way. Most people stop thinking when something works. When the car first plays a role in shopping for the household and makes it easier, the next time the owners of the car will choose to go by car again. Only with real and important changes, at tipping points, does clear space appear for bringing in new paradigms and for creating new habits.

This power of habits finalises the description of the 10 driving forces of frequent car use. We have split these forces along the line; more societal inspired – more inspired via individual motives. Other divisions are possible. For example, two driving forces are related to deliberate realising patterns and locations that are car-oriented; (spreading out of activities, highway locations), three driving forces are broad working (creating possibilities, no ethical boundary and habit) and at least five driving forces are socially and culturally focused; the urge for flexibility, convenience and comfort, community light, identity creation and feeling vulnerable and anxious.

These ten forces lead to frequent car use. They are intricately linked to our modern lifestyles, and to the individual reactions towards these lifestyles.

Each society creates stories about the ways to live together. These stories, *societal stories*, are grounded deeper than the driving forces and the stories condensate for some times the dominant thought patterns in a society. They record during that time the dominant wishes, expectations and anxieties in a society and they look quite stable, but will change over time, as history has shown. The rules of societal games in the Victorian age are now forgotten. With the normal low levels of reflection in most western societies actual societal stories of their period are seen as truth for most persons living in that period.

Related to car dependence, five stories seem dominant.

A story about manners

People in our societies are aware that their modern lifestyles create risks that are for a greater part 'man-made'. People know they are vulnerable, but they do not want to face that reality every day, so they build adjustments and mitigating arrangements. An important arrangement is the so-called 'community light'; having some reservations about fellow human beings but being friendly and polite and keeping a certain distance. With these manners it is difficult to ask neighbours, and sometimes even friends for real help. You need to help yourself and with this abstract 'doing it yourself' the car is a big help (for the core of this story, see Chapter 2).

A story about time

People are in modern societies expected to be flexible about new working conditions, on new ideas. To be able to satisfy all expectations and wishes a good income has to be brought into the household. Both partners have to work. All daily activities and care tasks still have to be done and this mostly in tight time frames, with explicit opening and closing hours. People have to combine tasks, and they can only meet all expectations by using the most flexible transport mode, the car (for the core of this story, see Chapter 3).

A story about education

In the past children could just be. Now, especially in middle class families, they have to become. Parents see raising children as an important task, and take this task seriously. There are many expectations around children. Because parents are anxious about the public space children are escorted to many locations. Their anxiety is related to stranger, danger and to traffic safety. The orientation of parents on risk protection and on interesting hobbies instead of playing in the neighbourhood make children less streetwise than in the past, and demands planning, for organising children's lives. The car is a great help (for the core of this story, see Chapter 2).

A story about identity

Living in our risk societies is based on abstract rules, functional responsibilities and rituals. We have to fit into these rules. Discovering your real self is not obvious. Authenticity is a state to reach. We want to be more than just our working life. We need to feel our strength, our emotions, our capabilities. We want to re-create, doing it ourselves, in our free time. For many people the car is an instrument, in the sense that cars can be instrumental in reaching the areas for authenticity and in the sense that car can create identities for their drivers (core of this story; Chapter 3).

A story about space

The different activities of modern life do not take place in one vicinity. Each activity has its own location and these locations have to be connected, to live a full life. The car is the great connector. And the highway becomes more and more a focus point for activities. Along the highway you can now work, live, go to a cinema. People grow up with highway experiences. Highways move from the periphery of our attention to the core (core of this story; Chapter 2).

All five societal stories need the car; please refer to Table 4.1 for details of this. One can conclude that the car has grown 'embedded' in our modern risk societies. Changing the role of the car means changing the current big societal stories.

Question 3: What is the current situation of car dependence and which persons and activities are more car dependent?

It is not very easy to quantify car dependence. The data make a clear quantification impossible and car dependence is partly a relative concept. Some people see themselves very fast as car dependent, while others, objectively sometimes even more car dependent, do not feel car dependent at all.

Most data on mobility are highly aggregated. For car dependence I had to work with estimations. In Chapter 4, I presented 'best guesses' and I made transparent how I reached the results.

For determining car dependence the question whether a journey can reasonably be made by alternative transport modes has to be answered. We looked for answers on this question in the literature. In general one can say that this is more difficult between 10 and 50 kilometres. Below 10 kilometres walking and cycling (and in cities tramways and buses) can be alternatives and above 50 kilometres the train becomes an effective possibility. The middle distances have rather inflexible public transport, making journeys by these modes twice as long as car journeys.

Also, persons living in the remoter rural areas, travelling with much luggage, and travelling in the evening, in the night or on Sunday mornings will be more car dependent. When many activities have to be combined car dependence increases and for people living and working along highways cars are, by far, the most efficient mode of transport.

We looked, for the Netherlands, more carefully at car-dependent activities. From a number of case studies (Jeekel, 2011) it could be concluded that for some activities car-dependence could be objectively measured. It is impossible to travel on weekdays between two Dutch cities at a distance of 30 kilometres or more at night without a car. By taking examples a list of reasonable car-dependent activities was made. This list contains; visiting friends at a distance in the evenings, going to cultural gatherings in the evenings, doing leisure activities in remote locations, or where much luggage is needed (for example mountaineering), the big weekly shop, going camping and caravanning. In general for these activities some 14.5 per cent of all car trips and some 16.5 per cent of all car kilometres can be defined as car dependent.

But the greater share of car dependence comes via car-dependent persons. Car-dependent persons are here defined as persons who are dependent on using their cars for at least 60 per cent of their car journeys. Six groups of persons can be presented as really car dependent.

First the *business drivers*: They do their work almost exclusively by car. They often work or meet at highway locations. Most of them have lease cars. Secondly, people who have to *combine tasks in daily life a lot*: This is the top of the pyramid of task-combiners. For many task-combiners, mostly women, the day is full of deadlines. Only the car offers them the flexibility to combine shopping, working, escorting and volunteer aid. A third group is *people living and working along highways*: A decade ago this group was non-existent but over the last decade many activities have been located along highways, and many housing areas are now to be found near highways. In some 10 per cent of Dutch households at least one person lives and work near the highway. The fourth group are *people living in rural areas*, especially in the remoter parts of rural areas with no bus services. Also car dependent are *commuters travelling to highway locations* for work. There is little public transport to be found on highways in most parts of Western Europe. To reach those locations a car is needed and the last group mostly do not drive themselves but have to be driven: *disabled people and less-able and care-needing persons.*

These six groups together contain, in the Netherlands, between 3.1 and 3.9 million adult persons, one third of all adults with cars. They are, however, not car dependent on all of their car journeys. Some 75 per cent of their journeys are car dependent. This means that on average 26 per cent of all car journeys and 28.5 per cent of all car distance is car dependent and is made by car dependent people.

In general we can conclude that *some 40 per cent of all journeys and some 45 per cent of all distance from car trips can be seen as car dependent. In 1995 we talked about 30 per cent and 35 per cent. This means that car-dependence is growing much faster than the growth in car mobility.* It becomes more difficult each year to make a trip with a mode other than the car.

A study from the UK came to similar conclusions. It was concluded that some 50 per cent of all car journeys were car dependent and a comprehensive study in Germany showed big differences between German cities on car-dependence. Differentiation in rates of car dependence can be seen and this gives room for well-designed policies. Unfortunately we have no studies on car dependence in other Western European countries.

Question 4: Which problems and perspectives related to frequent car use and car dependence can be noted in Western Europe in the near future?

Problems related to frequent car use and car-dependence may arise from at least two domains: oil, climate, and energy as the first domain and the role of the car in the fabric of society in Western Europe as the second domain.

The first domain: questions around oil, climate and energy. Here two sources of problems can be identified. The first source is fossil fuels. For some decades the car will still be dependent on fossil fuels. There is, however, a chance of scarcity of fossil fuels in the next 15 years. This scarcity may arise from geological barriers; the real exploitable supply of fossil fuels reaches a peak and stagnates on that peak. Scarcity related to smaller supplies than needed, created for geopolitical reasons by the fossil fuel producing countries, seems however more important. The demand on fossil fuels rises, for a large part related to economic development in countries of the developing world, and the supply will probably lag behind, as a consequence of decisions by oil producing countries. They note that their greatest clients, the OECD countries, are developing energy strategies for alternatives for their fossil fuels, and they want to profit for as long as possible from high demands.

Much is unclear about oil. The question is, for example, whether the strategy of the oil producing countries is purely economically based, or also finds a rationale in lower expectations of oil production in the longer run. Scarcity of fossil fuels will have at least three consequences. The first is higher petrol and diesel prices. The reaction of car drivers to higher fuel prices is inelastic. Drivers keep driving in the short term but in the longer term they make structural adjustments, like living nearer to work or buying a smaller car. A second consequence will be temporarily delivery problems. In these circumstances driving will be affected. The final consequence is structural delivery problems. Shell presents this consequence in the Energy scenario Scramble as a realistic option.

The other source has to do with the fight against global warming. To really stop global warming it now looks like, in 2050, emissions reductions in CO_2 of 60–80 per cent relative to the 1990 levels, are needed. In England this emission levels have been introduced by law. Transport will be late or very late in reaching these levels. There is still a huge growth in car mobility worldwide and there are few large scale and ambitious initiatives. This means that for the coming three decades other societal sectors will have to help the transport sector. Transport can grow slowly towards acceptable emissions because other societal sectors take more than their fair share of the burden. It is questionable how long other societal sectors will keep this form of solidarity and when they will start striving towards equity!

At this moment it looks like the transport sector has even difficulties in reaching the rather low emission targets for 2020. But also around CO_2 emissions, global warming and climate much is unclear. It can be questioned whether society at large will keep its attention on this problems. With economic crises sustainability problems often are seen as second row problems and it is questionable whether the international political circles will succeed in introducing these strong targets worldwide.

There is synergy between the two sources. Security in delivery of energy for cars and the fight against global warming both need a fast and clever shift towards far greater energy efficiency and towards new powertrains. The techniques are coming but the question will be whether consumers will change their purchasing behaviour. Far more energy efficient cars, lighter cars, alternative fuels, they all mean huge investments for producers, and these investments (not pilots!) are only made when producers are convinced that they can overcome the reluctance and the hesitations of their customers. Producers further need a long term strategy along with governments with strong, long-standing and determined goals.

Whether consumers, producers and governments can create a working triangle is doubtful. We will almost certainly be confronted with discontinuities in the coming decades. A real shift in the complete car fleet takes some 16 to 17 years, after having defined a complete business case. Actually, there is much attention on the electric car. This could be one of the possibilities for future driving but looking at the many problems that have to be solved first, electric cars cannot be seen as the '*deus ex machina*'. *From the literature it becomes clear that especially the period 2015–2030 may be a turbulent period and will be a crucial period for the future of car driving.*

The second domain is about car use in the social fabric of modern societies. First there are the effects that long-term are associated with frequent car use. Effects like noise, air pollution, use of public space, traffic safety, and harm to nature and landscape have been with us for more than 30 years. Via taxes the car drivers pay for the costs for mitigating and accommodating these effects.

Secondly there are consequences for the carless households, when society as a whole becomes more car dependent. What will be their position? As we saw most carless households are very young or rather old, are mostly single households, live more than average in cities, are far more than average without employment, and belong far more than average to the lower income groups. Very few families do not have a car but in single-parent families some 30 per cent are without cars. The profile of carless households shows, again on average, resemblance with the profile of the poorer strata of modern western risk societies. We did some research on mobility patterns of carless households in the Netherlands. Their mobility is 40 per cent lower than average and they walk, cycle and take public transport far more than average. A quarter of their kilometres travelled are by car, in 90 per cent of the kilometres as passengers.

We hardly know how carless households make the journeys that we defined as car dependent. Do they find alternatives or are these journeys not made by them? From literature we know that carless households can have problems in reaching hospitals, health care or cheaper supermarkets. They sometimes face difficulty in joining the possibilities and chances in modern society.

The same sometimes holds true for the poorer households. They can be faced with high percentages of their net incomes being necessary for driving. To be able to live in our modern societies especially in countries like Australia, the United States or Canada people need a household car, even when the burden on their budgets is big. Poorer households sometimes have to travel a long way to reach the jobs they are qualified for (see the Spatial Mismatch Hypothesis in the US), or see services and shops move out of their neighbourhoods. In the former welfare states of Western Europe there is little to no attention on these issues of potential or actual social exclusion related to mobility. Travel-on-demand schemes are seen as a solution.

Another potential risk of frequent car use is the loss of social cohesion. The car makes it possible to run away from the boundaries of territorial societal bonds. The car destroys the need of spatial vicinity. For car owners investing in their own neighbourhoods is no longer a necessity and becomes a matter of choice. Car drivers can also choose to invest in other networks, of friends and acquaintances at a greater distance. In modern neighbourhoods 'community light' is now the rule, indicating loose but friendly contact, with no possibilities to really ask for help or support. Many middle- and higher-income households especially choose to give their children a non-neighbourhood based education and they offer them a far wider spectrum of possibilities. The risk is that poorer households, or those without cars, will remain bound to neighbourhoods that will lose their former social qualities.

Driving in itself creates some risks. There is some concern about the growing costs of driving in household budgets. There is the feeling of insecurity related to car use. Obesity can be a greater problem in our western risk societies. There is stress and hurriedness related to the need for frequent car use and around 7 per cent of people with driving licenses experience some form of driving anxiety.

Most of the social problems mentioned here have not received much attention from decision makers. There is however one exception; congestion, essentially a problem of scarce road capacity at some hours, is a problem with a higher and middle class bias. Here we find the great commuters and the people involved in business to business services, a highly car oriented activity. Congestion can still grow in the coming decades, especially because of the still rather huge growth in freight traffic. Outside congestion many car users see the risks related to frequent car use as a sort of neutral problem related to their choice to drive.

Question 5: Is a form of governance on mobility necessary and which form could such governance take?

Car mobility will still grow and will get a somewhat higher share in the spectrum of transport modes. The dependence on the car will grow faster. At the same time there is a chance of scarcity in fossil fuels, and car mobility has to reach strong emission reductions. A cheap, reliable and clean car is needed. This car will probably arrive, however, the pace of change toward clean car mobility is rather slow and fuel efficiency and sustainability has been lost to the trend to bigger and more comfortable cars. With growing car dependence there is a chance that carless and poorer households will be marginalised. *This trend may lead, between 2015 and 2030, to great problems in car mobility.*

This question can be introduced as follows:

1. The car is essential to keep a differentiated society going; the car connects, binds and integrates in a scattered world. The car is indispensable in reaching the, in our modern societies considered normal, levels of flexibility which are very high. The car organises production and consumption and the car is a helper to be able to remain remaining consuming, an essential demand in our societies. Without a car people and households will miss chances and opportunities. This will be even more so in the future.

2. The car requires individual choices and individual driving behaviour. The sum of all this individual driving behaviour at societal level is scarcity and congestion, is a substantial contribution to exhaustion of fossil fuels, is environmental degradation, and is a substantial contribution to global warming. It looks like the boundaries for cars on fossil fuels are coming in sight, certainly when it will be seen from a world car growth perspective.

3. There is however as yet no winning technology for real and substantial new forms of car mobility, that can meet CO_2 and other sustainability objectives and have less or no dependence on fossil fuels.

In one sentence: '*the car is indispensable but may face an uncertain future*'. In the meantime car dependence is growing in all western societies! This means that our societies grow more dependent on an object and a system that will possibly not remain. This can be seen as 'looking for trouble'.

However, this is not yet a generally accepted idea. Many leaders in the debate on the future of car mobility believe that with great investment in car energy efficiency, with fast introduction of the newest vehicle technology and with a speedy shift towards electric cars we will be able to stay out of ugly problems regarding fossil fuels, we will be able to reach objectives on global warming, and we will be able to keep driving in a more modern and more sustainable way. In this book this is called the optimistic school.

There are, on the other hand, authors that believe that we are at the end of car mobility as we know it. The system that has led to frequent car use and to car dependence has now, in their opinion, so many anomalies that it cannot be expected to continue, even with great technological changes. These authors believe that all the technology combined will not do the trick to make a sustainable system of car mobility on the world scale. We will call this the realistic school, as these authors look at facts, figures instead of expectations and future perspectives.

The question remains whether a form of governance on car use will be needed. In Chapter 7 I introduced four scenarios for this governance. The central axis for governance is the increase in car mobility and car dependence in relation to the challenges on fossil fuels and climate. This deals with a fast introduction of cars not dependent on fossil fuels and with very low CO_2 emissions factors. An additional axis is the increase of car mobility and car dependence in relation to the perspectives of carless and poorer households. The affordability of driving and missing chances will become more important questions in more car dependent societies.

When a choice for governance is made the objectives to be reached have to be defined, but consensus fails here. Should the focus be on sustainability or on the maintenance of the system as we know it, with its related social arrangements? Or should we focus more on the perspectives of the more vulnerable households? *Car mobility can be seen as a 'wicked problem'.* The characteristic of wicked problems is the interrelation between problems and solutions. Government policies on wicked problems can be identified by disagreements from the immediate start of the policy.[2]

In these circumstances I have chosen to introduce four governance perspectives. These perspectives introduce basic attitudes to look at governance around car mobility and car dependence.

In the perspective *Following Wisely,* a technical orientation dominates. Car mobility is not seen from a normative perspective. Smaller interventions are prepared but governments are reluctant to make greater interventions. The credo is to wait for a bigger shift until the time is right.

In *Changing in Optimism* there is enough time. The scarcity on fossil fuels is not seen as an urgent problem and climate change is not seen as a big problem. Only minor targets for CO_2 emissions reductions are accepted. This scenario focuses on technical, especially IT-related changes. Until the change is absolutely necessary the actual car system has to be facilitated.

The focus in *Transport Transition* is on transition of the whole mobility system, towards sustainability. The existing regime of car mobility is unsustainable and needs to be changed by strong transition management. This perspective presupposes the possibility of strong societal steering.

In the last perspective *Creative Complexity* a transition toward a sustainable mobility system is also seen as necessary. However, one is doubtful of strong societal steering. In this perspective the already existing complexity and differentiation are seen as allies. Self-organisation of car drivers is seen as a fruitful road to travel.

In most Western European countries *Following Wisely* is the dominant perspective. This perspective can grow in quality by introducing some elements of the other perspectives:

First, the acknowledgement that there is a real chance of delivery problems with fossil fuels in the next 15 years, with a chance of reductions in car mobility. Second, more attention to the need of a fast shift towards a non-fossil fuel car fleet. Third, more focus on working in and from home. Fourth, a stronger focus

---

2 Fortunately there seems to be some common ground. The is consensus on the value of realising greater fuel efficiency, research on new power trains for cars, and on introducing more IT in cars.

on clever spatial planning, creating locations with easy access for other transport modes than cars. Finally, more attention to the self-organisation of car drivers, using complexity and differentiation as a tool, instead of as a burden.

And there are two additional elements:

- More attention on Western European demographics. Western Europe is growing older. Older people will remain driving but drive fewer than average car kilometres.
- Research on the conditions of the carless and the poorer households, with a focus on driving costs and on ways in which these households realise the car dependent trips.

With all these amendments a comprehensive strategy for the governance of car mobility can be developed. This strategy remains pragmatic in orientation.

9.2   A broader perspective

In this final section I will put frequent car use and car dependence in a broader societal scope. It is clear that in Western Europe we are on course for greater car dependence. For example, in the Netherlands within the next decade more than 50 per cent of all car journeys can reasonably not be made by another transport mode. We will become more dependent on car use in our functioning in modern societies.

In such situations of dependence you have to be able to trust the instrument that you are dependent upon. Here probably we will face difficulties in the next 15 years. The present modern car has had its time. It runs on fossil fuel, and creates levels of CO_2 emissions that are unsustainable. We have seen a development of modern cars towards comfort, weight and gadgets, thus using more energy than is acceptable. This development was the result of optimalisation in one direction. There are certainly aspects of path dependence and lock-in in our systems of car mobility. German researchers call this a situation of stagvation; stagnation in combination with innovation.

It is still rather unclear what will replace our fossil fuel-fed cars. There certainly are initiatives – like the hybrid car, the electric car, the hydrogen car – but none of these alternatives has such a momentum that we can speak of the winner. There is a chance that without a real, winning option, investments will be made in all these alternatives and the investments will probably be too small. This was exactly the situation over the last three decades. Only then, we still had our fossil fuel car winner!

At the moment of writing, early 2012, it does not look as if, in Western Europe, we will be able to find an inspiring route out of our locked-in position. This does not mean there are no inspiring initiatives. Frankly speaking, there are lots of inspiring initiatives but they are all too small and too little interrelated to form the basis for a big shift from fossil fuel cars to real sustainable car mobility for the whole world. *We also identified in the domain of car mobility a rather low governance capacity.* The network of all relevant organisations, all stakeholders on car mobility at the national levels, but also at the European level, has too little capacity to act jointly and collectively in a preferred direction.

Converting our Western European car fleet to a sustainable car park will from past experience take some 16 years, after we have a real business case. This means that this sustainable car fleet will be with us in Western Europe at the earliest around 2032. Speeding up is possible, but one has to take into account that the decision to purchase a new car is taken at the household level. It will be a heavy job for governments to influence the households in their countries in such a way that they will decide earlier than expected on buying a new ands far more sustainable car. As this is a function of household budgets it is clear that economic and financial development of our Western European economies will be a crucial factor in the speed of converting the car fleets!

The question is whether we will actually have this period until 2032 to make the shift. As seen in Chapter 5 there are possibilities that scarcity on fossil fuels will be with us earlier than this. We will possibly need to reduce emissions from our car before 2030 with some 40 per cent (basic year 2000). The conversion of the car fleet will be very difficult, will demand a sense of direction and huge investment. Meanwhile the dependence on cars in our societies will just be growing further. This means that *in Western Europe societal arrangements will become more dependent on a product that, in itself, will be in an uncertain and still unclear transition process.*

One strategy could be looking at the uncertainties, to organise things so that our car dependence will not increase further, at least not until we know what future sustainable car mobility with no or far less fossil fuels will really look like. This is a form of precaution.

However, this will not be easy to achieve. Because cars tell our major societal stories the trend is towards greater dependence and we will have to change directions in the driving forces for car use.

Putting the increase of car dependence on hold will mean adjustments in the big societal stories that were introduced in Chapter 4. A few possible directions for change can be given.

In the story on manners the wish for 'community light' could change into organising resilience by creating common goals in neighbourhoods. Leaving each other alone costs energy and creates feelings of anxiety. The public space, the space for everybody, has to be maintained and cherished again.

The story on time asks for relaxation. Flexibility is good, but has probably gone too far. We should be able to relax time frames. And we can change our way of looking at work times. Results are far more important than attendance. Working at home and teleworking need stimulation. We now focus on clock time and forget our biological clock or our personal time experiences, our living time. This demands change.

In the story on educating children we could shift the focus on 'becoming' towards 'being'. Children are happier when they see their parents and can relate to them. Countries that have kept a good balance between working and spending time with children have happier children, as concluded in a recent UNICEF report and parents should look at the real figures on crime instead of being victims of their media-driven perceptions.

The story on identity asks for authenticity. What exactly is authentic about driving your car to the walking machine in the fitness centre? A leisurely walk is more useful. What is authentic in a wilderness experience along with your favourite singer on the car stereo? Authenticity is experiencing the weather in all its variety!

In the story on space the trend towards highway locations should be reversed. Restructuring the existing areas would be wiser than creating new business and housing areas while letting older areas lose their qualities.

In societies with these societal stories there will be less car dependence. Alongside this, investments need to be made to raise the quality of life and hence accessibility for carless and poorer households. This looks like more rewarding investments than trying to tackle congestion. Congestion is just the result of our lifestyles, so changing lifestyles would help best!

To conclude: If we really want to decrease the growth in car dependence we should at least decrease the following trends:

- task-combining in ever tighter time frames;
- realising ever more highway locations;
- avoiding perceived risks by escorting children everywhere;
- concentrating leisure activities in car-dependent locations;
- and going everywhere by car without thinking.

Addendum
Facts and Figures on Car Mobility in Europe

In this Addendum the focus is on facts and figures on car mobility in Europe, with a special emphasis on Western Europe. For this perspective I define Western Europe as containing 13 bigger countries: Finland, Sweden, Norway, Denmark, Germany, Austria, Switzerland, France, Belgium, the Netherlands, the United Kingdom and Ireland. Of these, seven have more-or-less comprehensive and more-or-less actual National Travel Surveys. I will concentrate on these – Sweden (2005 data), Germany (2008), Switzerland (2005), France (2008), Flanders (2008, the Northern part of Belgium), the Netherlands (2009) and the United Kingdom (2009)[1] – and sometimes broaden to the whole of Western Europe, or to the whole of Europe.

In Part 1 a short historical perspective will be presented. How did car mobility rise to its, now indisputable, supremacy in modern mobility? After this perspective, different themes will be introduced as presented below:

Car ownership, car growth, driving license, cost of driving	2
General mobility growth, car mobility growth	3
Number of trips, distances travelled, modal splits in general	4
Motives for travel, number of trips, distance, transport modes used	5
Focus on work motives, family affairs and escorting, and leisure, related to car use	6

---

1 Data sources are:
   • *Sweden*: RES 2005 – 2006 The National Travel Survey, SIKA (Swedish Institute for Transport and Communications Analysis), 2006
   • *Germany*: Mobilitat in Deutschland 2008 : Ergebnisbericht; Struktur – Aufkommen – Emissionen _ Trends, Infas und DLR, Berlin, 2010
   • *Switzerland*: Mobilitat in der Schweiz; Ergebnisse des Mikrozensus 2005 zum Verkehrsverhalten, Bundesamt fur Statistik / Bundesamt fur Raumentwicklung, Neuchatel, 2007
   • *France*: La mobilité des Francais. Panorama issu sur1' enquete nationale transports et dep[lacements 2008, Commissariat General du Developpement Durable, 2010
   • *Flanders*: Onderzoek Verplaatsingsgedrag Vlaanderen 3 (2007–2008), Universiteit van Hasselt, Instituut voor Mobiliteit, D. Janssen e.a., 2008
   • *Netherlands*: Mobiliteitsonderzoek Nederland 2009, Tabellenboek, Rijkswaterstaat, 2010
   • *United Kingdom*: Transport Trends, 2009 edition and National Travel Survey: 2009, Department of Transport, 2010

This Addendum ends with the presentation, as a summary, of the General Western European Mobility Patterns (GWEMP); what constitutes at the moment and at a rather abstract level, mobility in Western Europe?

1.   Car mobility: A short historical perspective

Before the Second World War, car density in Europe was rather low, certainly when compared to the US. In *Shaping Transport Policy* (2011, 91) Filarski presents an interesting table showing the car density in the US in 1938 (195 cars per 1,000 inhabitants) compared to car densities in Europe, between 42 (UK and France) and 11 cars (Netherlands) per 1,000 inhabitants. Car ownership in Europe started its take off between 1959 and 1963, with a huge growth between 1965 and 1985. Thanks to continuing, growing prosperity between 1950 and 1980, from 1965 to 1985 the middle classes in Western Europe were successively able to afford cars. After 1985 car growth did continue in most Western European countries, but at a slower pace.

The introduction of the car in Western European societies was an interesting process. In the Netherlands this can be split in four phases.[2]

The first runs from the days of the motor-car pioneers to 1923. During this period car ownership started growing, primarily among the well-to-do households. In particular, men with an interest in technology and speed liked cars and during this phase buyers had to be able to repair their cars themselves. Around 1920 car use became less exclusive and in the Netherlands the first acceleration in growth can be seen between 1918 and 1923.

The second phase runs from 1923 to 1957 and it was during the period to 1938 that a real car fleet came into existence. The period between the wars was essential for the development of the car society and in most Western European countries a start was made on what eventually became national highway systems. Although very few households owned cars, people without cars became acquainted with the essentials of speed without rails by using the new mode of transport, the bus.

During the Second World War the car fleet diminished. Superior cars were confiscated by the occupying forces. In 1939 barely 15 per cent of all Dutch cars were over eight years old, compared to 76 per cent in 1946. Growth in car ownership continued until 1957 but in these first after post-war years this took place alongside a new transport mode, the moped. These mopeds were bought by the lower and middle classes who could not yet afford cars.

The third phase runs from 1957 to 1980. After 1957, and in the Netherlands more clearly after 1963, wages rose, and car ownership came into view for the middle classes, followed somewhat later by the highest-earning lower classes. In 1960 in the Netherlands for the first time more kilometres were covered by car

---

2   These phases are described in the thesis of Staal, *Automobilisme in Nederland; Een geschiedenis van gebruik, misbruik en nut.* (Staal, 2003) Staal uses a theory on diffusion of products.

than by public transport. In 1963 the 50 per cent barrier was crossed; more than 50 per cent of all kilometres were covered by cars. In the sixties cars were primarily bought for business purposes, and a little later for leisure. Commuting started in the early seventies, following the development of suburbanization.

The Netherlands had a rather slow diffusion of the car in society. Most Western European countries followed the same phases, but a little faster, and the role of bicycles in Dutch society must be mentioned as a braking mechanism.

The last phase began 1980. Between 1985 and 1995 mobility growth in Dutch society was, on average, 2.5 per cent a year and car mobility growth was a little higher. The years 1994 and 1995 are the starting point for further statistical analysis.

2.   Aspects of car ownership

Car ownership has grown in recent decades, and is still growing. In most Western European countries, there are now more households with two or more cars than households without a car. From Table A.1 it can be concluded that 18 to 25 per cent of all households have no car. Forty-three to 55 per cent of all households have one car. And 20.0 to 32 per cent of all households own two cars. More than two cars can be found in 2.6 per cent to 6.0 per cent of all households.

The rule of thumb for the selected seven countries is 21 per cent carless households, 49 per cent one car households, and 30 per cent households with two or more cars.

Table A.1   **Ownership of cars, by households,[3] selected countries (%)**

Country	No car	1 car	2 cars	More than 2 cars
Germany	18.0	53.0	23.0	4.0
United Kingdom	25.0	43.0	26.0	6.0
France	19.0	45.0	32.0	4.0
Switzerland	18.8	50.6	25.1	5.4
Flanders	18.2	53.6	24.7	3.5
Sweden	25.0	52.0	20.0	3.0
Netherlands	20.9	54.9	21.6	2.6

Source: Different National Travel Surveys.

Car ownership is still growing quickly in Western Europe. In the Netherlands car ownership in 1995 was nearly 5.7 million cars, compared to 7.6 million cars in

---

   3   In each table and in the text of this Addendum I use the figures of the country from the year of their National Travel Survey (2005, 2007, 2008 or 2009). I present the source page.

2009, a growth of 33.3 per cent or around 2.4 per cent yearly. Car density in the Netherlands went from 390 cars per 1,000 inhabitants in 1995 to 460 cars per 1,000 inhabitants in 2009. In the UK between 1994 and 2008 the number of cars went from 23 million cars to 31 million cars, a growth of nearly 35 per cent, or around 2.3 per cent yearly.

In France from 1994 to 2008 the number of cars grew from 26.4 million to 32.7 million, a growth of 24 per cent, and around 1.7 per cent yearly.

The number of cars per 1,000 inhabitants is called car density and data for the whole of Europe are available (Eurostat, 2009).

Starting with Western Europe, many countries had in 2008 a car density of around 460–500 cars; Sweden (466), Norway (464), UK (464), Belgium (481), France (485). The German-speaking countries have somewhat higher densities; Germany (503), Austria (515), Switzerland (530) and especially Luxemburg (681!). When we look at the southern part of Europe we see Spain with 489 and Italy with 609. Further to the east we see mostly lower densities: Poland (422), Estonia (412). Latvia (411), Czech Republic (426), Slovakia (286), Romania (187), Bulgaria (310), Hungary (333) and Croatia (350). Only Lithuania (496) and Slovenia (520) have high car densities.

The conclusion from these patterns of car growth and car density should be that Western Europe and Southern Europe will soon reach the situation where a car is available for every second person in the population. But which households have two cars, one car, or no car? In order to answer this we must take a quick look at the statistics in a few selected countries.

In Germany for example over half of the households with a net monthly income lower than 1,500 euros do not own cars, and households with net incomes higher than 3,000 euros have in majority more than one car. Above 6,000 euros net income, 25 per cent of households have three or more cars (Abbildung 3.34, 58).

When being asked for reasons not having a car, 50 per cent of the non-car owners considered a car too expensive, 19 per cent gave health- or age-related reasons, 16 per cent did not need a car, and 5 per cent rejected cars (Abbildung 3.35, 59). Carless households are more to be found in the cities Berlin (41 per cent), Bremen and Hamburg, more in former Eastern Germany (around 25 per cent) and among singles (below 30 years 38 per cent, and above 60 years 46 per cent). Twenty-five per cent of single parent families do not own a car (Abbildung 3.38, 62).

In the UK the level of 25 per cent carless households has been stable since 2004. More than 43 per cent of the households in the lowest two income quintiles had no car. The majority of households in the highest two quintiles had two cars or more (62). Of the non-car owners 33 per cent said they were not interested in driving, and 32 per cent could fall back on family or acquaintances. Cost factors have become more important in recent years, health and age reasons were important for 19 per cent. Carless households are to be found in the London boroughs (38 per cent) and in other major urban areas (31 per cent on average). Convenience was by far the most important reason for having more than one car (Att, Fig. 2) and in rural areas more than half of all households had two or more cars.

In France income is also important, but the location of the households matters even more. The highest car ownership rates are to be found in the rural and peri-urban areas, with very low rates in the bigger cities and extremely low rates in Paris (graphique 2, 104). The number of students, inactive people, unemployed and older people without a car is above average and 25 per cent of single parent families do not own a car. Carless households live on average far nearer to railway stations (104).

In the Netherlands 43.5 per cent of single households (singles are a third of all households) do not own a car. In families with two children only 4 per cent do not own a car and a little over 50 per cent of these families now have more than two cars.

Related to car ownership is the availability of a driving licence. In the Netherlands 80.4 per cent of the people over 18 hold a driving licence. This figure is somewhat higher than that for car ownership.

In our seven countries the possession of driving licences is on average 81.0 per cent. Non-car ownership is, as we saw, 21.0 per cent. This means that in some 10 per cent of carless households somebody can drive. Among men the average driving license density is 87 per cent, among women 74.5 per cent. The highest density in driving licenses is in Germany, the lowest in the UK and Sweden, the countries with the highest percentage of carless households. The greatest gap between men and women is to be found in Flanders.

Table A.2   Driving licence density

Country	Male %	Female %
Germany	93.0	83.0
United Kingdom	81.0	66.0
France	87.0	80.0
Switzerland	89.2	74.4
Flanders	89.9	73.8
Sweden	82.0	70.0
Netherlands	86.9	75.5

Source: Different National Travel Surveys.

The price of driving

It can be seen from the statistics that owning a car is often too expensive for households at the lower end of the income spectrum. But how expensive is driving in relation to the household budgets in modern western European households?

The price of driving is not easy to calculate. We have material available from at least four countries; the Netherlands, Germany, Austria and the UK.

Starting with the Netherlands: The Dutch drivers association *ANWB* has an insightful table for all relevant prices related to car ownership and car use, the Autokostentabel (ANWB, 2011). For a smaller to middle class car the ANWB shows figures of – everything included; buying, depreciation, taxes, maintenance, insurance, petrol – between 4,600 and 5,400 euro yearly, 390 to 450 euros monthly.[4] In the Netherlands the average annual distance by car is 13,000 km. This means that in the Netherlands 1 km driving costs between 35 and 40 eurocents. Cars are also leased (8 per cent), and employers subsidize up to a maximum of 19 cents per km for work-related purposes (in the Netherlands almost 50 per cent of the annual driven kilometres per car are for work, see later). This means that in most households prices for driving will drop by an average 25 per cent, this part being paid by the employers.

The average net household income in the Netherlands[5] is 2,600 euros per month. Average car use in car-owning households is 17,000 km per year (related to 1.3 cars per household). This means that an average household spends 130 per cent of 35 to 40 cents, minus 25 per cent (employers' costs) on driving costs. This is 370 to 430 euros per month and this is between *14.5 per cent and 16.5 per cent of the average net household income.*

The figures from the *United Kingdom* are comparable. *Transport Trends 2009* (trend 2.6b) states that the overall cost of motoring has dropped below its 1980 level. This had to do with the relative costs of purchasing a vehicle. Petrol and oil, taxes, insurance went up, but the average disposable income more than doubled between 1980 and 2009. Households do spend more on driving, because they have more cars in their households. Driving accounted for 14 per cent of the total of British household expenditure.

For *Germany* IFMO presented results in *Mobilitat; Der Einfluss von Einkommen, Mobilitatskosten und Demografie* (2007). For all households the cost of driving per km in 2003 was 36 eurocents (IFMO, 2007, 68). For a comparable 2,600 euro net-income household (also with 1.3 cars per month) the figure was 428 euros per month comparable with 16.5 per cents.

For *Austria* we have also recent figures from *AEC: Euromobil; Vergleich von Mobilitatskosten in ausgewahlten Landern der EU* (2009, 26–27). Seventeen per cent of the yearly net household budget in Austria is spent on driving costs.

Although comparison is not easy it can be stated that *the cost of driving for average households in Western Europe seems to be mostly around 15 per cent (plus/minus) of the net household income.*

---

4 There is a big differentiation. It is difficult to arrive at less than 280 euros per month, but for example more luxurious cars in the *Autokostentabel* stand for 800 euros and above (see also Jeekel, 2011, 16).

5 In the Netherlands often a 'one and a half' income situation can be seen. A man works full-time, his wife or partner, part-time.

3. Mobility and car mobility in general

In this section the development of mobility will be analysed. However some problems exist, initially with the national travel surveys. Each country presents its statistics using different classifications. More importantly some countries only look at adults, while others take the whole population over six years old into account. In a number of countries there is a distinction between daily mobility (in France for example journeys up to 80 km) and longer travel. Others do not make that distinction or present, as the Netherlands, no figures on holiday travel. Taking note of these differences I will present the statistics here, but with some reluctance.

The growth of *general mobility* in kilometres or miles can essentially take place in three ways:

- The first is through population growth. More inhabitants will, average distance per person remaining stable, lead to higher mobility figures.
- The second is the growth in travel distance by each member of the population. The number of inhabitants being stable, this will lead to higher mobility figures.
- The third, only relevant for car miles/km driven (and not for passenger miles/km driven), is by diminishing car occupancy. Lower car occupancy means that more cars have to drive for the same distance.

Table A.3   Distance travelled per person, per car, and transport and population growth

Country	Distance travelled per person		Distance per car, recent years	Transport growth; population growth 1995–2009
	around 1998	around 2008		
Germany	40 km (1998)	41 km (2008)	14.300	7% ; 2%
United Kingdom	34.5 km (2000)	34.1 km (2009)	13.200	11% ; 5%
France	23 km (only daily mobility, 2000)	25.1 km (2008)	13.000 (113)	12% ; 6%
Switzerland	38.1 (2000)	38.2 (2005)	13.900	14% ; 5%
Netherlands	34.8 (2000)	35.1 (2009)	13.000	12% ; 6%

Source: Different National Travel Surveys.

In all six countries it is clear that growth in distance travelled per person is now rather slow or non-existent. There are however differences. With Table A.3 in mind we can now draw some conclusions.

First on *general mobility*: In Germany, France, and the Netherlands the distance travelled per person is still growing in the first decade of the new millennium, although far slower than between 1995 and 2000. In the UK (with a diminishing distance travelled per person between 2000 and 2009!) and in Switzerland mobility growth is only the result of population growth.

On *car mobility* Europe wide figures are available: Eurostat (2009, 103) presented in Panorama of Transport 2009 (2009, 103) the distance growth in km. of passenger cars (drivers and passengers). The general figure for the older 15 EU countries was 1.7 per cent per year between 1990 and 2006, and only 1.3 per cent between 1995 and 2006. For all 27 EU countries from 1995 to 2006 the car mobility growth was 1.6 per cent. This means that the new EU Member States had rather high growth rates.

For the 13 western European countries car mobility growth between 1995 and 2006 was on average 15 per cent, for Belgium 12 per cent, for the Netherlands 13 per cent, for France 13 per cent, for Germany 7 per cent, for Denmark 10 per cent, for Norway 19 per cent, for Sweden 12 per cent, for Finland 24 per cent (!), for the United Kingdom 12 per cent, for Switzerland 16 per cent and for Austria 17 per cent.

For the new Member States this growth rate was far greater between 1995 and 2006. For example Bulgaria 61 per cent, Czech Republic 28 per cent, Cyprus 43 per cent, Latvia 82 per cent, Lithuania 106 per cent (!) and Poland 77 per cent. The conclusion should be that car mobility is still growing in Western Europe, but at a far slower pace than in the new member states.

A short case study; mobility growth in the Netherlands

In a short case study I will present the pattern of mobility growth in the Netherlands (Jeekel, 2011). As can be seen from Table A.4 there is still some mobility growth. Between 1995 and 2000 this growth was 10.6 billion km, yearly on average 2.1 billion. Between 2000 and 2009 this growth was 11.8 billion km, yearly on average 1.4 billion. The mobility growth is thus slowing down.

Table A.4   Total distances passenger mobility in the Netherlands, 1995–2007

Year	Car drivers (km)	Car passengers (km)	Car in total, % of total transport distance	Public Transport (train, tram, bus), km	Cycling (km)	Walking and other transport modes (km)	Passenger kilometres total (km)
1995	80.1	51.3	74.7	21.0	14.7	8.9	176.0 / 100
1996	81.4	51.3	75.1	21.0	13.9	9.0	176.7 / 100
1997	83.1	53.4	74.6	22.2	14.9	9.6	183.2 / 103
1998	85.0	52.1	75.0	22.7	14.0	9.0	182.8 / 103
1999	88.4	52.9	75.7	22.5	14.1	8.6	186.6 / 106
2000	89.1	52.0	75.6	22.9	14.1	8.5	186.6 / 106
2001	90.2	51.4	75.5	23.1	14.0	8.8	187.6 / 107
2002	91.9	52.3	76.2	22.7	13.9	8.6	189.3 / 108
2003	92.9	53.2	76.6	21.1	14.8	9.0	190.9 / 108
2004	95.4	56.2	77.0	22.4	14.4	8.6	196.9 / 112
2005	94.9	53.6	76.5	20.6	15.4	9.3	194.0 / 110
2006	95.8	52.2	75.9	22.0	15.0	10.0	195.1 / 111
2007	97.5	52.0	75.8	21.5	15.1	11.2	197.2 / 112
2009	99.0	49.9	75.1	23.7	15.1	9.7	198.4 / 13

Source: Mobiliteitsonderzoek Nederland 2009.

We can define the nature of this growth by looking at some developments between 1995 and 2009.

From Table A.5 a few conclusions can be drawn. The number of cars has grown faster than the kilometres driven by cars. This means that the number of kilometres per car is diminishing. Passenger kilometres are growing more slowly than car kilometres (13.2 per cent vs 23.9 per cent) which means that car occupation rates have diminished; more people drive alone in their cars.

Table A.5   Different increases related to passenger mobility, the Netherlands 1995–2009

Increase in number of cars	33.3%
Increase in car kilometres (only car drivers)	23.9%
Increase in kilometres on highways	29.0%
Increase in passenger kilometres (car drivers and car passengers)	13.2%
Increase in Dutch population	5.8%
Increase in total mobility growth	13.0%
Increase in people with driving licence	12.6%

Source: Mobiliteitsonderzoek Nederland 2009.

The increase in kilometres covered on the Dutch highway system is greater than the increase in car kilometres, meaning that more traffic is taking the highway, compared to other roads. The total mobility growth is more or less comparable with the growth in passenger kilometres, meaning that car mobility in the Netherlands is the driving force behind mobility growth.

In Mobiliteitsbalans 2009 the Kennisinstituut voor Mobiliteitsbeleid (KiM) presented the elements of Dutch car growth:

First the period between 1985 and 2008 – 50 per cent of the car mobility growth is related to mobility for work. Of this growth 45 per cent is related to more people going to work (work participation of women grew quickly in the 1980s and early 1990s) and 55 per cent to longer commuting journeys. Thirty per cent of total car mobility growth is related to leisure (people travel more often and further away and both account for half of the explanation) and the last 20 per cent is related to population growth.

For the period 2000–2008 another figure was presented. Twenty per cent is related to population growth and 80 per cent to work related mobility (30 per cent longer commuting journeys and 50 per cent extra labour participation in particular by women).

In conclusion, for the Netherlands important factors behind the growth in general mobility are the growth in distance travelled per person, plus population growth. Where growth in car mobility is concerned mobility related to work has been most important, with women joining the labour force, and with growing commuting distances.

4.   Journeys, distances and modal splits, with a focus on car mobility

Table A.6   Journeys, distances and modal splits in selected countries

Country	Journeys per day	car	PT	bike	walk	Distance per day	car	PT	bike	walk
Germany	3.5	58 (43+15)	9	10	24	41 km	79 (54+25)	15	3	3
United Kingdom	2.9	63 (41+22)	11	2	23	34 km	79 (50+29)	13	1	3
France	3.15	65	8	4	23	31 km	83	14	1	2
Switzerland	3.3	40	15	5	39	37.3 km	67*	20.5	4	6
Flanders	3.15	65 (47+18)	5.5	13.5	14	41.6 km	74	10	4	2
Sweden	3.1	50	12	9	24	40 km	78	14	3	3
Netherlands	3.0	48 (33+15)	5	27	19	35.1 km	74.5	13	8	2

Source: Different National Travel Surveys.

From Table A6 different elements can be highlighted. First the number of *journeys:* for all seven countries between 2.9 and 3.5, and on average 3.25 per day. Most of these journeys are made by cars, but there are huge differences. The spectrum runs from 48 to 65 per cent, on average 56–57 per cent, with something like 39 per cent drivers and 17.5 per cent passengers.[6] The number of public transport journeys is rather low in the Low Countries (concurrence with biking) and is in selected Western European countries around 10 per cent, with a high score for Switzerland. The slow modes are responsible for between 25 and 44 per cent of the journeys, on average 34–35 per cent, with two thirds walking (with an exception in the Low Countries).

Next the *distance:* on average around 36 kilometres per day per person. The greater part of the distance is travelled by car, ranging from 74 per cent to 83 per cent,[7] meaning on average somewhere around 78–79 per cent. Public transport, accounting for 10 to 15 per cent, is 12–13 per cent (with currently a real exception of Switzerland). The slow modes are between 3 and 11 per cent, on average 7 per cent, somewhat more cycling than walking.

Looking at the individual countries and concentrating on journeys made by car, interesting differences and patterns can be seen.

In *Germany* 3.5 journeys are made each day. Fifty-eight per cent of these are made by car, 43 per cent by drivers, 15 per cent by passengers. In the three-city *Länder* of Berlin, Hamburg and Bremen only 42 per cent of all journeys are made by car. On the other hand, in Saarland 71 per cent of all journeys are made by car. Older singles make only 39 per cent of their trips by car, their use of slow modes is 50 per cent. Younger families and middle-aged couples make 64 per cent of their journeys by car. Looking at income we see real differences, from 30 per cent car journeys for those with the lowest incomes, to above 60 per cent for those with the highest incomes. Where there are three or more cars in the households 73 per cent of the journeys are made by car (67/68). The most car mobile age is between 40 and 49 years with 67 per cent journeys by car. Seventy-nine per cent of all distance travelled is done by car (54 per cent drivers and 25 per cent passengers). This figure is constant during the last decade. The spectrum goes from 71 per cent in the bigger cities, to per cent in the rural areas. Per day 41 kilometres are travelled.

In the *United Kingdom* 2.9 journeys are made each day. Sixty-three per cent of all these journeys are made by car (41 per cent driver, 22 per cent passenger). The number of trips by car, by men, has fallen by 20 per cent since 1996 and the number of trips for women has grown by 9 per cent in that same period. Ten per cent fewer trips are made overall when compared with 1996 (especially fewer car passenger trips) but the total time spend on these has remained more or less the same, showing that journeys are getting longer. Until their sixties women make more journeys than men, especially between the ages of 30 and 50. The overall

---

6   Acknowledging that the percentage for Switzerland is too low.

7   With again, the unusual exception of Switzerland.

distance travelled by car in the United Kingdom is comparable with Germany; 79 per cent, 50 per cent drivers and 29 per cent passengers. Higher income groups travel 2.5 times more km than the lowest income groups.

In *France* 3.15 journeys are made and 65 per cent of these are made by car. The highest number of journeys is made by those between 25 and 45 years of age, nearly 4.0. In Paris only 12 per cent of all journeys are made by car, compared to 85 per cent in peri-urban suburbs. Eighty-three per cent of distance travelled is by car. The daily distance travelled is 25.4 kilometres.[8] Daily travel distance within Paris is very low, 14 km, while in the peri-urban region it is almost 38 km. Active people between 20 and 50 years of age travel on average the longest distance daily, nearly 36 kilometres.

In *Flanders* 3.15 journeys per day are made, of which 65 per cent are by car (47 per cent as driver, 18 per cent as passenger). Women make fewer journeys than men (3.0 vs 3.3). More journeys are made than average, when the income is higher than average. The distance travelled per person per day is 41.6 km. (men: 49 km, women; 43 km) 74 per cent of this distance is covered by car.

Finally, in the *Netherlands* 3.0 journeys are made each day. Forty-eight per cent of all journeys are made by car, 33 per cent by drivers and 15 per cent by passengers. In general people with the highest incomes make more journeys; 15 per cent more, as households with the lowest incomes make 15 per cent fewer. Thirty-two per cent of journeys by the lowest income groups are by car, compared to 57 by the highest income groups. Women between 30 and 50 years of age make the highest number of journeys; 22 per cent more than average, meaning they make more journeys than men of the same age. Men between 50 and 60 years old are the most car-mobile; 61 per cent of their journeys are by car. Of all journeys within the major cities 36 per cent are made by car (nearly half by slow modes). Of the journeys in rural areas 54 per cent are made by car.

Seventy-five per cent of all distance is travelled by car. People with lower education travel only 55 per cent of the average distance, while people in the highest education categories travel 60 per cent above that average. Better educated people travel nearly three times as much as lower educated people. Higher income groups travel twice as much as lower income groups. Over distances between 20 and 50 kilometres car use accounts for more than 80 per cent of the distance. It is in journeys of over 50 km that the influence of the train as an alternative mode is seen. People from major cities travel 60 per cent of the distance by car, in rural areas this is more than 82 per cent.

From these results is noticeable that although there is in the selected countries a difference in car use when looking at journeys (48 per cent to 65 per cent by car), that difference cannot be seen when looking at distance (mostly somewhere around 77–79 per cent by car). *This means that longer distances are being travelled by*

---

8   In France, mobility statistics are split between daily travel and longer distance travel (including holidays). This explains the low average distance.

car. For shorter distances other modes may be chosen. The differences in trips can be explained by looking at whether the geography of a country is suitable for walking or cycling.

Another point to note is the dependence of the scale of car traffic on income. *Higher income, longer distances, more car traffic, seems to be the rule* and the number of car journeys and car distance covered goes up with lower population density. *Higher densities; less car use, is the rule.*

Finally, the role of women: In most countries they make fewer journeys but between 30 and 50 years of age in the United Kingdom and the Netherlands their average number of journeys is higher than that of men in the same age group. They travel however shorter distances, an indicator of some 'rushing around behaviour'.

The broader picture: Car mobility in Europe

Until now the orientation has been on Western Europe and more specifically on the seven countries of Western Europe with well-defined and actual National Travel Surveys. Now a broader picture will be presented. In *Panorama of Transport* (2009) the Statistical service of the EU, *Eurostat*, presented the average daily distance per mode of travel for all European countries. The data presented here is for cars and public transport (separate bus and coach, railways, trams and metro) but not for slow modes[9] (See Table A.7).

---

9 This means that in most countries in their Travel Surveys car percentages of 78 per cent with small modes end above 80 per cent in the Eurostat data without the small modes.

Table A.7   European countries: Car distance travelled per day in relation to car % in modal split cars/public transport

Car mobility	Car share in modal split above 80%	Car share between 75 and 80%	Car share below 75%
More than 30 km per day	Iceland Norway Finland Slovenia Italy France Luxembourg United Kingdom Lithuania	Switzerland	
Between 25 and 30 km per day	Germany Sweden	Belgium Denmark	
Between 20 and 25 km per day		Netherlands Austria Spain	Estonia
Less than 20 km per day	Portugal	Latvia Poland	Bulgaria Hungary Romania Slovakia Czech Republic

Source: Edited from the figures 4.69 en 4.70, Eurostat 2009.

Although the basic data differ somewhat from the National Travel Surveys a rather well-defined picture can be seen. Most Western European countries are in the left and higher part of the table. These countries are high on car use and high on kilometres travelled per day. Most Eastern European countries (exceptions being Slovenia and Lithuania) are in the right and lower part of this table. These countries are not yet strong on car use and lower on kilometres travelled per day.

5.   Motives for mobility

Motives in general

Why are people mobile? For what reasons do they travel? Comparing the motives for mobility is not very easy. Each country has its own classification. A general classification on the basis of the seven National Travel Surveys has been made. On the motives a rather clear picture arises (see Table A.8).[10]

Starting with *journeys*; it is clear that (except for France) *leisure leads to the greatest number of journeys.* In general the figures are; work-related 23 per cent, shopping 20 per cent, leisure 33.5 per cent. Other motives present 23.5 per cent together.

With *distance* the pattern is somewhat different. Work-related is 32 per cent, shopping-related 10 per cent, and leisure-related 42 per cent. The other motives present 16 per cent together.

This means that work journeys are the longest (ratio trips–distance is 1.4), shopping trips the shortest (ratio 0.5), and leisure journeys fall in-between, but rather long (ratio 1.26).[11]

Table A.8   Motives for general mobility, trips and distances, seven selected countries

Country		Work	Business	Work together	Education	Escorting	Shopping	Personal	Leisure	Other
Germany	trips	14	7	**21**	6	8	**21**	12	**32**	
	distance	21	12	*33*	4	6	*9*	8	*40*	
UK	trips	15	3	**18**	5	6	**20**	20	**31**	
	distance	19	8	*27*	2	2	*12*	14	*42*	1
France	trips	21	6	**27**	10	12	**19**	8	**24**	
	distance									
Switzerland	trips	23	3	**26**	8	2	**20**		**41**	3
	distance	23	9	*32*	4	1	*11*		*45*	7
Flanders	trips	15	6	**21**	7		**21**	5	**36**	10
	distance	21	14	*35*	4		*10*	3	*41*	7
Sweden	trips			**30**			**20**	10	**33**	6
	distance									
Netherlands	trips	17	3	**20**	9	10	**19**	4	**37**	1
	distance	28	8	*36*	5	5	*9*	3	*42*	

Note: In **bold**: most important data.
Source: Different National Travel Surveys.

10   Although we miss France and Sweden on distance, and have questions about the data on journeys in Switzerland.

11   Leisure trips are longer in Germany and the UK, work trips are rather long in the Low Countries.

Car use by the motives

For most countries we have information on journeys being undertaken by car. For work purposes around 70 per cent of journeys are made by car, for education a wide range exists, with an average around 30 per cent; for shopping around 55 per cent and for leisure around 52 per cent.

Table A.9   Journeys by motive, percentage of all journeys made by car, selected countries

Country	Trips made by car			
	Work	**Education**	**Shopping**	**Leisure**
Germany (2008, 43, 65, 76, 93)	70	27	58	49
United Kingdom (2009, 71)	75	42	63	
France (2008, 72)	75	38	67	60
Flanders (2008, 66)	70	31	44	48
Netherlands (2009, 28)	60	16	43	47

Source: Different National Travel surveys.

On distances we have other figures, but only for three countries. For work respectively 66.5 per cent (Switzerland), 75 per cent (Flanders) and 79 per cent (Netherlands) of the distance is travelled by car, on average 75 per cent.

For shopping respectively 85 per cent, 80 per cent and 75 per cent of the distance is travelled by car, on average 80 per cent.

For leisure respectively 69 per cent, 73 per cent and 80 per cent of the distance is by car, on average 75 per cent.

What we can state is that for work the car is the favourite mode of transport, while for shopping, and for leisure, the car is used primarily for greater distances. Shorter trips for these motives are made by other modes.

Some differentiation can be noted. In *Germany,* for example, there is a difference between household types. Singles and couples between 30 and 60 years of age make more work-related than leisure journeys. Escorting children is, after leisure, the second most important category for families with children under the age of six, and for single parent families. For youngsters education is, after leisure, the second most important journey category. For older people shopping is the most important category after leisure.

Looking at the journeys car drivers make, 34 per cent are for work-related reasons, 22 per cent for both leisure and shopping. There are not many car passengers on car journeys for work purposes (15 per cent) and many for leisure (44 per cent).

6. Car use and activities

For three motive categories some extra insights will be presented on car mobility and car use.

Working and car use

In Western Europe work related journeys account for 23 per cent of the total. These journeys are, together, responsible for 32 per cent of the total distance travelled and 75 per cent of this total distance is travelled by car. Journeys for work purposes are mostly made alone. Of all car use between 30 and 38 per cent of passenger kilometres is for work-related purposes. Over 90 per cent of the journeys for business purposes are made by car (Jeekel, 2011, 30).

In the Netherlands work-related purposes were responsible for the greater part of the growth in mobility in recent decades. More people, especially women, are going out to work, are driving to their workplaces, and the distances between home and work have grown. For the Netherlands the *Mobiliteitsbalans 2007* (KiM, 2007) contains an overview of relevant work related developments.

First for housing: Between 1996 and 2003 over 700,000 houses were built in the Netherlands, of which a substantial number are located adjacent to cities but near highways. The expectation was that these new housing areas would form new elements in the existing urban web. However, being so near to the highway system this expectation was not met. People living in these areas see themselves from a mobility point of view as being rather 'footloose' and orient towards different cities along the highways, thus becoming car-dependent (KiM, 2007, 128).

Work facilities have also tended to be situated near highways. Between 1996 and 2002 the growth of work locations was the strongest near airports (32 per cent), near highway access roads (27 per cent) and near the bigger railway stations (12 per cent). As we will see later 40 per cent of Dutch employment opportunities are now to be found near highways.

This all fits in a more general trend. In the past the commuting patterns were from the suburbs to central cities, in relative heavy traffic streams. These streams made it possible to offer proper public transport. Over the last thirty years work has decentralised, along with housing. In Western Europe the standard commuting pattern has become a criss-cross pattern. Such a pattern creates difficulties for public transport providers. For example, in Germany people were being asked how easily they could reach their work location by different transport modes. It was concluded that their work locations could easily be reached by private car (89 per cent), but with greater difficulty by bicycle (54 per cent) or by public transport (55 per cent).

In *France* commuting distances grew by 22 per cent between 1994 and 2008. The longest are to be found in the peri-urban areas around Paris. The growth in distance travelled is greatest in the peri-urban areas outside the Paris region and the longest travel times are to be found in the Paris region. The car is used in 72.3 per cent of work-related trips.

An interesting polarisation can be seen. Car use is diminishing in Paris, in the banlieue of Paris and in other major densely populated city regions. For example: in 1994, 20 per cent of work-related journeys within Paris were made by cars, in 2008 only 13 per cent. On the other hand, in the peri-urban and rural areas car use is still growing. The highest car use for work-related journeys is to be found in the peri-urban areas outside Ile de France; 93 per cent.

In Flanders the car is used for intermediate distances from home to work. Below 6 kilometres alternatives are still important (walking and cycling), whereas from 25 km the train gets a higher share (above 15 per cent) of the work-related trips.

In the Netherlands and in Switzerland a study has been made of gender issues related to work travel. Most women in the Netherlands work part-time (78 per cent), most men work full-time (78 per cent). Part-time workers travel on average 32 km for work-related reasons, full-time workers 41 km. Women between 25 and 60 years of age on average travel – for work related reasons – 69 per cent of the distance travelled by men (Jeekel, 2011, 31). In Switzerland the average length of a working journey is 9.9 km for women and 13.0 km for men: this is 76 per cent of the distance travelled by men.

Mobility for work is related to education and income. In the Netherlands people with a higher level of education travel 2.5 times the average distance for work and the highest income groups travel 3 times the average distance (Jeekel, 2011, 32).

Teleworking and working at home can be seen as alternatives to commuting. But how important is teleworking? We have figures for three countries. In the UK in 2006 3 per cent of those who were employed always worked from home, and a further 5 per cent did so at least one day per week (National Travel Survey, 2006, 49). For another 10 per cent of the employees it was possible to work from home. In the Netherlands in 2003 between 3 and 22 per cent of the employees worked from home with some regularity (more than once in two weeks), on average 6 per cent (SCP, 2003, 89–114). Avoiding longer commute distances is an important motive; the average distance to work of regular home workers is longer than normal (SCP, 2003, 106). Higher education and work in the public sector are indicators for working at home/teleworking and in Sweden in 2005 1 per cent of those in employment teleworked, while 21 per cent of the employed stated they had work tasks appropriate for teleworking (41).

From these data a picture arises; more than one fifth of the employed population can telework relatively easily, half of this group actually works at home sometimes and half of this last group (4–8 per cent) work regularly at home.[12]

Family affairs, escorting and car use

Modern western risk societies have a huge variety of household types. At the moment in the Netherlands more single households exist than families and the same is the situation in Germany.

In Germany in 2006 the figures were: 38.8 per cent single households, 30.0 per cent couples and 31.2 per cent families.

In the Netherlands in 2003 there were 34 per cent single households, 29 per cent couples and 30 per cent families, plus 6 per cent single parent families.

Table A.10   Types of households in the Netherlands, 1990–2020 (%)

	1990	2003	2020
Single household	29	34	39
Couple	28	29	28
Family with children	36	30	26
Single parent family	5	6	6
other	1	1	1
Average household	2.44 persons	2.28	2.15

Source: Sociaal Cultureel Planbureau (2005).

In 2006 the *De tijd als Spiegel; hoe de Nederlanders hun tijd besteden* (SCP, 2006a) was published. This publication gives a broad overview of the life of modern households. People between 20 and 65 years of age needed from 2000 to 2005 one hour extra per day for all their duties (SCP, 2006a, 13). More people are combining working with running daily life, caring, escorting children or family members. 'Task combiners' is a category in Dutch social statistics which describes those people who, in a week, spend at least 12 hours working and at least another 12 hours caring and housekeeping. Forty per cent of the Dutch population is now seen as a task combiner, and this percentage is growing (SCP, 2006a, 22). Combining tasks often leads to difficult chain journeys – from house to work to school to shops to house – and, as Harms (2003, 44) showed, to a greater necessity to use a car. We already noticed that women in their middle years (30 to 50) make

---

12   Cairns et al. (2004, 345) stated that in the UK with a proper telework and home work strategy a 12 per cent reduction in car use for commuting could be the result.

more journeys than men, but travel fewer kilometres. These chain journeys form an important part in their daily mobility.

The Dutch Social Research Institute (*Sociaal Cultureel Planbureau*, SCP) sees two working partners, children and task combining as a rather stressful situation (SCP, 2006a, 23) but single households also have to 'task combine' a lot. Couples in general are under somewhat less pressure to perform task combining. Women in the Netherlands cannot reduce their housekeeping and caring activities further; these activities take up twice as much of their time as their male partners. After recent growth the time that Dutch men spend on caring and housekeeping has stagnated. The SCP figures show that in the Dutch family life, with most households getting together for dinner each day forces women, having many obligations and duties for caring and housekeeping, to continue working on family affairs in the evenings.[13]

In the mobility statistics of other countries little can be found that directly relates to car use for family affairs and housekeeping. However, most have a statistical category that the Dutch mobility statistics misses. This is the category *escorting*. Escorting by definition does not cover all the aforementioned activities but does include caring for the most part. Escorting means driving people, often children or the elderly, to clubs, friends, hospitals. Note that the school run is not completely covered by this category. The school run is mostly an element in the motive 'education'.

For Switzerland, the figures show that in 2005 94 per cent of escorting was done by car and women between 30 and 50 years of age dominated in this motive. For the Netherlands it looks like some 15 per cent of the journeys made by women of these ages are for escorting (MON 2009, 8.11 category *overig* [miscellaneous]). For most of the seven selected countries we have more escorting figures.

Table A.11   Data on escorting, selected countries

Country	Trips (of 100%)	Distance (of 100%)	By car, distance
Germany	8.1	5.5	75%
United Kingdom	20*	14	84%
France	13		
Flanders	11.6	8.3	84%

Note: * In the United Kingdom the category is called escorting and personal business. It seems difficult to split this category. This explains the rather high figure.

Source: Different National Travel Surveys.

---

13   In *Changing Rhythms of Family Life* (2006) Bianchi states that combining tasks leads to diminishing leisure and rest time of women.

The British Travel Survey 2006 states, 'while younger women make more escort trips than younger men, men aged 50 and over made more escort trips than women in the same age group.' Including both escort education trips and other escort trips, women aged 30–39 made over 25 per cent of their trips escorting someone else and in Germany, for parents with children aged under 6 escorting makes up 26 per cent of their trips, lessening to 12 per cent when children are over 6.

From the data the picture arises that escorting is an important motive; somewhat like 11 per cent in journeys and 8 per cent in distance, with a peak for women between 30 and 50 years of age with children, where escorting accounts for 25 per cent or more of their journeys. Escorting is very car-dependent; it looks like more than 80 per cent of the distance for escorting is travelled by car, making escorting one of most car-oriented motives. Note that the second highest mode is walking, not public transport. Public transport has no role in escorting.

Escorting and the school run take time. Most travel for these motives is not on highways, but on smaller and slower roads, mostly in built up areas. And there is another element; escorting and the school run have to be done at fixed times, within fixed time frames. This creates some stress, especially when people combine tasks. For example, *De tijd als spiegel* shows the following: In 2005 compared to 2000 computer use grew (from 0.5 hours a week to 2.5 hours a week) and more and longer working by women was an important trend (SCP 2006a, 48). In the Netherlands the dominant work pattern remained. However, especially in families, the SCP concluded that subjective time pressure was a factor. People were asked whether they felt hurried and that certainly was the case.[14] People recorded that they felt hurried three times a week, 'the fuss that exists to keep on fixed times for dinner, for work, for school in a very active life, with circumstances that you cannot control is not shown in official statistics (free from SCP, 2006, 25)'. In a more recent report *Tijd op Orde?* (2010) the SCP concluded that 60 per cent of the women and 52 per cent of the men felt pressured a number of times during a week and this pressure was concentrated between the ages of 30 and 50. Combining tasks and escorting played their role, 'Being active in different domains of life means not only an accumulation of activities, but also coping with different expectations and duties, in the different domains. One has to connect and to change gear all the time' (SCP, 2010, 42).

Also important in this complexity are the opening hours of shops, services and offices. Time ordering still finds its basis, at least in the Netherlands, in collective rhythms. Most shops are not open in the evening and the same is true for government services, or for technical professional services. People are expected to visit or use these services during the day, when they are at work or have escorting duties. This relative rigidness is a problem, especially for women. The car is a big help under these circumstances of task combining, time pressures and fixed opening hours.

---

14   Figures for Flanders: Moens (2004).

Car use and leisure

Leisure is – looking at the number of journeys – the most important motive for using a car. Although with a smaller distance than the second motive (work-related journeys) it remains the most important motive looking at the distance driven (driver plus passengers). When the focus is only on kilometres driven leisure loses its first place. This has to do with car occupancy rates. For work purposes the car mostly only contains the driver, for other motives, there are usually one or more passengers.

Leisure is a complex motive, composed of visiting friends and relatives, eating out, sports, entertainment, making walking and cycling journeys, and leisure day journeys.

The complex character of this motive creates problems, as the different sub-motives are not presented in the official mobility statistics. Not being able to differentiate means that we only know that the combined leisure activities form a mobility pattern. That pattern is consistent in each of the seven countries; elder people make relatively more journeys, and travel greater distances, the working population is below the average. Workdays accounted for around 25 per cent of all leisure journeys and this goes up on Saturdays to around 45 per cent and on Sundays to about 70 per cent. On longer leisure journeys the car dominates, for shorter trips we see a focus on walking.

In *Overwegend Onderweg* Harms clarified the leisure journeys with statistical data from 2004 (see Table A.12).

Social leisure is about visiting friends and relatives, plus some caring tasks. These trips are on average somewhat longer than the others, 62 per cent of the journeys, and 87 per cent of the distance are by car.

Recreational leisure is more about walking, cycling, sight-seeing. Trips are on average shorter and most involve walking and cycling. For visiting, more than 75 per cent of the journeys is by car.

Table A.12   Journeys and distances in leisure, the Netherlands, 2004

	% of journeys	Distance involved
Social	39	48
Recreational	20	14
Sport	12	7
Horeca	5	4
Culture	2	2
Miscellaneous	22	25

Source: Harms, Overwegend Onderweg, SCP, Den Haag (2008).

Sport, eating and drinking and cultural activities; although the activities differ they have in common a somewhat smaller distance and a greater use of slower transport modes.

The category 'miscellaneous' is about visiting libraries or churches, taking part in hobbies and voluntary work; 55 per cent of the longer journeys and 80 per cent of the distance travelled is by car.

This division of the leisure motive clarifies the situation that just over 50 per cent of the journeys is made by car, but 80 per cent of the distance. There are a great number of shorter journeys for leisure such as going to the library, walking, cycling, going to the pub, visiting nearby friends etc. Leisure is not differentiated geographically; more-or-less the same amount of journeys and distance is made from different locations (for example, see Germany; 3.20, 43).

For the social leisure motive extra information is available from the Netherlands. In the *Netherlands Kinship Panel* (Mulder and Kalmijn, 2004, Table 4.4) it is shown that family members live on average 34 kilometres from each other. As education levels rise this distance grows. This means that travelling to your family takes time, more time than in the past, when family members lived close by. Dutchmen travel three times a week for social visits, not only to family members.

In most western risk societies a general trend in leisure can be seen. Free time has been diminishing since the nineties, but at the same time this free time is spend more dynamically, and more consumption, experience and adventure oriented. To end with a vision of the future,

De ruimtelijke ontwikkelingen zullen de commercialisering, intensivering en individualisering van de vrije tijd faciliteren en versterken. De verwachting is een verdere toename van het aantal en de variëteit van vrijetijdsvoorzieningen, een voortzetting van de schaalvergroting en clustering met andere functies en toenemende vestiging aan de rand van stedelijke gebieden (de opkomst van pleasure peripheries). Dit enorme aanbod van voorzieningen heeft een aanzuigende werking op de consument. Alleen al het feit dat er meer te doen en te beleven is dan voorheen, verleidt de consument ertoe vaker onderweg te zijn.
(Harms, 2008, 238)

Spatial developments will facilitate and strengthen the commercialisation, individualisation and intensification of leisure time. One can expect a further growth in number and variety of leisure facilities and services, a growth in scale and a clustering with other functions. More leisure activities will be located at city frontiers, the growth of the pleasure peripheries. This greater supply will lead to more consumption. There is more to do, and more to see than in the past, and consumers will travel to these possibilities (author's translation).

7.   Towards a General Western European Mobility Pattern (GWEMP)

Looking at the figures for the seven selected countries many similarities around mobility and car mobility can be seen. As a summary, a first draft of the General Western European Mobility pattern will be introduced, on two levels; the general and the household.

General level

Mobility in Western Europe is growing slowly. Growth comes mostly from growth in population, and less from growth in distance travelled per person. This last form of growth is diminishing rapidly. Within the total mobility (globally 36 kilometres per person per day) it is only the car kilometres that are still growing. Car kilometres account for 79 per cent of the total kilometres travelled and this share will grow over the next decade to over 80 per cent. Public transport has now 12.5 per cent of the distance travelled, and the slow modes account for 7.5 per cent, somewhat more cycling than walking. The car share in journeys (globally 3.15 mobility journeys per person day) is lower, around 56.5 per cent, public transport covers 10 per cent of the journeys, and 34.5 per cent are made by slow modes. Shorter trips are made walking or cycling and for journeys over 7 km the car dominates. Although the distance travelled is not growing much in Western Europe, the number of cars still is still growing relatively quickly. While distance travelled per car is diminishing, the number of cars per household is rising. *Cars become individualised consumption articles,* like clothing. Also diminishing is the number of passengers per car. Driving often means driving alone.

Household level

Nearly half of Western European households have one car. Thirty per cent have two or more cars and this share is growing, and 21 per cent of the households – mostly single households, the elderly, and the very young households – do not own a car. The majority of families with children now own two or more cars. Driving license density (81 per cent, men 87 per cent, women 74.5 per cent) is a little higher than car density (79 per cent). In one tenth of the households without a car someone can drive.

Car driving costs 15 per cent of net household income. This share is more or less constant. Poorer households have fewer cars and travel smaller distances in general. The same holds true for less educated households. Reliance on a car is highest in rural and in peri-urban areas.

Car mobility is least important for households in Western Europe's major cities, where public transport has an important share. The car dominates in work related mobility (with 70 per cent of the journeys and 75 per cent of the distance). Most work travel is driving alone. For leisure and shopping the car is used primarily for longer distances. And education is the least car dependent motive. Because cars have more passengers in leisure traffic this motive is the winner in total car distance travelled. An important newcomer among the motives is escorting, an indicator that the car is becoming a necessity in the fulfilment of the duties of daily life.

Western European pattern in the whole of Europe

It is interesting to see this emerging Western European pattern in relation to the total European pattern. Southern Europe still has a somewhat higher mobility growth, but the big difference is with the new EU Member States in the eastern part of Europe. Here real mobility growth, and within this mobility growth almost exclusively car growth, can be seen. Here the car is still in an emancipation phase, comparable with the period 1975–1985 in Western Europe.

References

Aarts, H., and Dijksterhuis, A. 2000. Habits as knowledge structures: Automaticity in goal-directed behaviour. *Journal of Personality and Social Psychology*, 78(1), 53–63

AASHTO, 2008. Primer on transportation and climate change, Washington

Achterhuis, H. 1998. Mobiliteit tussen schaarste en rechtvaardigheid, in *Cultuur en Mobiliteit*, edited by H. Achterhuis, and B. Elzen. Den Haag: Sdu

Adams, J. 1999. The social implications of hypermobility: Speculations about the social consequences of the OECD Scenarios for EST and BAU Projections, ENV/EPOC/PPC/T (99)3/FINAL/REV1

Adams, J. 2005. Hypermobility: a challenge to governance, in *New modes of Governance: Developing an Integrated Approach to Science, Technology, Risk and the Environment*, edited by C. Lyall, C. and F. Tait. Ashgate: Aldershot

Adelman, J. 2004. The real oil problem. *Regulation*, 27(1), 16–21

Adriaansens, S. and Hendrickx, J. 2008. Choices in threefold. Home maintenance and improvement between household production, informal outsourcing and the formal economy, Brussels, HUB University College Research Paper 2008/29

AEC. 2009. Euromobil; Vergleich von Mobilitatskosten in Ausgewahlten Landern der EU,

Aigle, T. and Marz, L. 2007. Automobilitat und Innovation. Versuch einer interdisziplinare Systematisierung, Berlin, *WZB – Discussion paper SP III 2007-102*

Aigle, T. et al. 2007. Mobil statt fossil. Evaluationen, Strategien und Visionen einer neuen Mobilitat, Berlin, *WZB discussion paper SP III 2007-106*

Ajzen, I. 2002. Residual effects of past on later behaviour: Habituation and reasoned perspectives. *Personality and Social Psychology Review*, 6(2), 107–22

Akerman, J. and Hojer, M. 2005. How much transport can the climate stand? – Sweden on a sustainable path in 2050. *Energy Policy*, 34, 1944–55

Al Husseini, M. 2006. The debate over Hubbert's Peak: a review. *Geo Arabia*, 11(2), 181–210

Aleklett, K. 2008. Peak oil and the evolving strategies of oil importing and exporting countries; Facing the hard truth about an import decline for the OECD countries. Paris, *ITF/OECD; Oil Dependence, Round Table* 139, 40–96

Altheide, D. 2002. *Creating Fear: News and the Construction of Crisis*. New York: Aldine de Gruyter

Annema, J.A., Hoen, A. and Geilenkirchen, G. 2007. Review beleidsdiscussie CO2 – emissiereductie bij personenvervoer over de weg; Achtergrondnotitie voor de Raad voor V&W, de Algemene Energieraad en de VROM-Raad, KiM en MNP

ANWB, no date. Autokostensite. Available at: www.anwb.nl/kiezen-en-kopen

Armitage, J. 1999. Paul Virilio: An Introduction. *Theory, Culture & Society*, 16(5–6), 1–23

Aronczyk, M. 2008. Taking the SUV to a place it has never been before; SUV ads and the consumption of nature. *Invisible Culture, an electronic journal for visible culture, 9*

Ascher, F. 2006. *Le movement dans les societes hypermodernes*. Vanves, CERIMES.

Augé, M. 1995. *Non-places: Introduction to an Anthropology of Supermodernity*. London: Verso

Avelino, F. et al. 2007a. An interdisciplinary perspective on Dutch mobility governance, *Erasmus University paper*

Avelino, F. 2007b. Newspeak or paradigm shift? How to interpret the Dutch discourse on 'Transition to Sustainable Mobility', paper presented to the conference on Interpretation in Policy Analysis, Amsterdam

Axhausen, K., Urry, J. and Larsen, J. 2007. The Network Society and the networked traveller, in *Travel Demand Management and Users: Pricing, Success, Failure and Feasibility*, edited by W. Saleh and G. Sammer, Aldershot: Ashgate, 89–109

Bachiri, N. 2006. L'etalement urbain et la mobilité quotidienne d'adolescentes de territoires urbain de la Communauté Metropolitaine de Quebec. These Ecole d'Architecture, Quebec, Université Laval

Bachiri, N., Despres, C. and Vachon, G. 2008. Fighting teenagers sedentarity; the challenges of mobility in exurbia. *Medio Ambiente y Compartamiento Humano*, 9, 47–67

Baker, K. and Kaul, B. 2002. Impact of changes in household composition on home improvement decisions. *Real Estate Economics*, 30(4), 551–66

Bakker, P. and Hal, van J. 2007. Understanding travel behaviour of people with a travel-impeding handicap: Each trip counts, paper for TRB Annual Meeting, Washington

Balaker, T. 2003. The quiet success: Telecommuting impact on transportation and beyond. Los Angeles: Reason Foundation

Balaker, T. 2006. Why mobility matters, Los Angeles: Reason Foundation

Balkmar, D. 2007. Men, cars and dangerous driving: affordances and driver – car interaction from a gender perspective, Paper, Past Present, Future Conference, Umea, 14–17 June

Bamberg, S., Ajzen, I. and Schmidt, P. 2003. Choice of travel mode in the theory of planned behaviour: The roles of past behaviour, habit, and reasoned action. *Basic and Applied Social Psychology*, 25(3), 175–87

Bauman, Z. 2001. *Liquid Modernity*. Cambridge: Polity Press

Bauman, Z. 2006. *Liquid Fear*. Cambridge: Polity Press

Beck, U. 1992. *Risk Society; Towards a New Modernity*. Nottingham, UK: Sage Publications

Beck, U. 2001. *Die Modernisierung der Moderne*. Frankfurt am Main: Suhrkamp Taschenbuch Verlag

Beck, U. 2002. The cosmopolitan society and its enemies. *Theory, Culture & Society*, 19(1–2), 17–44

Beck, U. 2008. Risk society's 'Cosmopolitan Moment', Lecture at Harvard University, 12 November

Beck, U., Bonss, W. and Lau, C. 2003. The theory of reflexive modernisation: problematic, hypotheses and research programme. *Theory, Culture & Society*, 20(2) 1–33

Beck, U. and Sznaider, N. 2006. Unpacking cosmopolitanism for the social sciences; a research agenda. *British Journal of Sociology*, 57(1), 1–23

Becker, U., Clarus, E. and Friedemann, J. 2009. Klimaschutz im Verkehr – Paradigmenwechsel. *Wissenschaftlichen Zeitschrift der TU Dresden*, 132–6

Beckmann, J. 2001. Heavy Traffic: Paradoxes of a Modernity Mobility Nexus, in *Mobility and Transport: An Anthology*, edited by L. Drewes-Nielsen and H. Oldrop. Copenhagen: The Danish Transport Council

Beckmann, J. 2002. Sustainable transport and reflexive mobility. Paper, International seminar, Managing the Fundamental Drivers of Transport Demand

Beckmann, J. 2004. Mobility and safety. *Theory, Culture & Society*, 21(4/5), 81–101

Béland, D. 2005. The political construction of collective insecurity: From moral panic to blame avoidance and organized irresponsibility. *Centre for European Studies, University of Calgary, Working Paper Series 126*

Berg, J. van den. 2005. *Omgaan met rij-angst*. Bohn, Stafleu en Van Loghum

Bergman, S. and Sager, T. 2008. The ethics of mobility, in *Rethinking Place, Exclusion, Freedom and Environment*, edited by S. Bergman and T. Sager. Aldershot: Ashgate.

Bianchi, S. 2006. Changing rhythms of family life. Paper, in Time Use and Gender seminar, London, 14 June

Bianco, M. and Lawson, C. 1996. Trip chaining, childcare and personal safety. *Women's Travel Issues, Proceedings from the Second National Conference, Washington.*

Blumenberg, E. and Waller, M. 2003. The long journey to work: Federal transportation policy for working families. Washington DC: The Brookings Institution

Bohm, C. et al. 2006. *Against Automobility*. Oxford: Blackwell.

Bollier, D. 2006. The growth of the commons paradigm, in *Understanding Knowledge as a Commons*, edited by C. Hess and E. Ostrom. Cambridge: MIT Press, 27–41

Boomkens, R. 2006. De nieuwe wanorde en de stad, *Enneus Heerma-lezing*

Boon, W., Geurs, K. and Wee, van B. 2003. Sociale effecten van verkeer: een overzicht. Paper voor Colloquim Vervoersplanologisch Speurwerk

Borden, I. 2005. Drive: Urban experience and the automobile. *Critical Architecture Theory, Public Lecture Transcripts*

Boutellier, H. 2002; De Veiligheidsutopie: Hedendaags onbehagen en verlangen rond misdaad en straf. Den Haag: Boom Uitgevers

Boutellier, H. 2006. Fataal Vitaal: de criminologie van een vloeibare samenleving. *Justitiële Verkenningen*, 32(5), 27–44

Bowden, C. and Moseley, M. 2006. The quality and accessibility of services in rural England: A survey of the perspectives of disadvantaged residents, Wolverhampton, ADAS

Breedveld, K. 1999. 4+2=7: arbeidsduurverkorting, taakcombinatie, opvoeding en onderwijs van 4 tot 12 jaar oude kinderen. Den Haag: SCP studie in opdracht van de commissie Dagindeling

Brindle, B. 2003. Kicking the habit: some musings over the meaning of 'car dependence'. *Road and Transport Research*, 12(3) 61–73; (4), 34–40

Brink, Van der, G. 2004. Hoger, Harder, Sneller ... en de prijs die men daarvoor betaalt, in *Bijdragen aan WRR rapport Waarden en Normen*, Den Haag: Wetenschappelijke Raad voor het Regeringsbeleid

Brodersen, S. 2003. Do-It-Yourself in North Western Europe: Maintenance and improvement of homes. Copenhagen: Rockwool Foundation Research Unit.

Brons, M., Nijkamp, P., Pels, E. en Rietveld, P. 2006. A meta-analysis of the Price Elasticity of Gasoline Demand. A System of Equations Approach. Den Haag: *Tinbergen Institute Discussion Paper*

Brookings Institution. 2010. The suburbanization of poverty: Trends in metropolitan America 2000–2008, Washington DC

Bruggink, J. 2005. *The Next 50 Years: Four European Energy Futures.* Petten: ECN

Buchan, K. 2008. A low carbon transport policy for the UK, Phase 2, Final Report, www.transportclimate.org, last accessed 23-9-2010.

Buhn, A., Karmann, A. and Schneider, F. 2007. Size and development of the shadow economy and of do-it-yourself activities in Germany. *CESIFO Working Paper No. 2021*

Bundeambt for Statistik/Bundesambt fur Raumentwicklung. 2007. Mobilitat in der Schweiz: Ergebnisse des Mikrozensus 2005 zum Verkehrsverhalten, Neuchatel

Bureau Louter. 2007. Ontwikkelingen en prognoses regionale kantorenmarkt, Delft

Burbridge, S., Goulias, K. and Kim, T-G. 2005. Travel behaviour comparisons of active living and inactive living life. Paper presented at 84th Annual Transportation Research Board Meeting, January 9–13

Burke, B. 2001. Hardin revisited: A critical look at the Perception and the Logic of the Commons. *Human Ecology*, 29(4) 449–76

Cairns, S. et al. 2004. Smarter choices: Changing the way we travel. Department for Transport UK, London

Campbell, C. no date. Oil Depletion – The Heart of the Matter. Paper for The Association for the Study of Peak Oil and Gas

Campbell, S. and Park, Y. 2008. Social implications of mobile telephony: The rise of the personal communication society. *Sociology Compass*, 2(2), 371–87

Canzler, W. and Marz, L. 1996. Festgefahren? Der Automobilpakt im 21. Jahrhundert. Berlin, *WZB Discussion Paper FS II* 96–108

Canzler, W., Knie, A. und Marz, L. 2006. Osten ergluht, China ist jung … China als Katalysator einer postfossilen Mobilitatskultur? *Informationen zur Raumentwicklung*, nr 8, 439–46

Canzler, W., Kaufmann, V. and Kesselring, S. 2008. *Tracing Mobilities: Towards a Cosmopolitan Perspective*, Aldershot: Ashgate

CapGemini i.s.m. Transumo. 2009. Trends in Mobiliteit 2009

Carribine, E. and Longhurst, B. 2002. Consuming the car: Anticipation, use and meaning in contemporary youth culture. *Sociological Review*, 50, 181–96

Carrigan, M. and Szmigin, I. 2004. Revisiting the construct of convenience in consumer research: The paradox of convenience consumption. Paper 4th, International Critical Management Studies Conference, 4–6 July, Cambridge, UK

Castells, M. 2000. *The Information Age, Volumes I, II and III.* Oxford: Oxford University Press

CBS 2000. *Onderzoek Verplaatsingsrepertoire Korte Ritten.* Voorburg/Heerlen

Cebulla, A. 2008. Adjusting to a more modern world: Have risk perceptions changed between generations ? *SCARR* Paper 21/2008

Chappels, H. and Shove, E. 2004. Comfort: A review of philosophies and paradigms. Centre for Science Studies, Lancaster University, UK

Chea, L. et al. 2007. Factor of two: Halving the fuel consumption of new U.S. automobiles by 2035. Cambridge, MA, MIT Laboratory for Energy and Environment

Chlond, B. and Ottman, P. 2007. Das Mobilitatverhalten Alleinerziehenden und ihre Aktivitaten ausser Haus. *Deutsche Zeitschrift fur Kommunalwissenschaften*, 46(2), 46–61

Choplin, A. and Delage, M. 2011. Mobilites et espaces de vie des etudiants de l Ést francilien; des proximites et dependances a negocier. *Cybergeo European Journal of Geography*, 1 July

Commissariat de Developpement Durable. 2010. France ; La mobilité des Francais. Panorama issu sur l'enquete nationale transports et deplacements 2008, Paris

Conley, J. and Tigar McLaren, A. 2008. *Car Troubles*. Aldershot: Ashgate

Cortright, J. 2008. Driven to the brink: How the gas price spike popped the housing bubble and devaluated the suburbs. CEO's for Cities, May

Coutard, O., Dupuy, G. and Fol, S. 2002. La pauvreté péri-urbaine; dépendance locale ou dependence automobile ? *Espaces et Societes*, 108–9; 155–76

Craig, L. 2005. How do they do it? A time-diary analysis of how working mothers find time for their kids. *Paper University of New South Wales, Social Policy Research Centre*

Cremers, M., Backera, V. and Faun, H. 2007. Afstand tot werk of afstand tot de arbeidsmarkt; een onderzoek naar ruimtelijke mobiliteit van lager opgeleiden in Twente en Zuid Limburg, Maastricht, E 'til

Creswell, T. 2008. Understanding mobility holistically, in *The Ethics of Mobility*, edited by S. Bergman and T. Sager. Aldershot: Ashgate, 129–40

Crozet, Y. 2008. Mobilité durable: des inflexions aux ruptures, quelles politiques publiques? *TEC*, 198, 3–12

Cuenot, F. 2007. Is the automobile industry able to reach 120 gr CO_2/km average on new cars sold by 2012 with no more regulation from the EU? *Paper presented at the ETC Congress*

Currie, G. et al. 2009. Investigating links between transport disadvantage, social exclusion and well-being in Melbourne – preliminary results. *Transport Policy*, 16, 97–105

Dant, T. 2004b. Car care project, the professional care and maintenance of the private car

Dargay, J., Gately, D. and Sommer, J. 2006. Vehicle ownership and income growth, worldwide 1960–2030, *Energy*, 28, 1–7

Dargay, J. and Hivert, L. 2005. The dynamics of car ownership in EU countries: A comparison based on the European Household Panel Survey. Paper presented at the European Transport Conference, Strasbourg

Davey, A. 2004. Coping without a car. Paper prepared for the New Zealand Institute for Research on Aging

De Beer, P. 2007. How individualised are the Dutch? *Current Sociology*, 55(3), 289–313

Dejoux, V. and Armoogum, J. 2010. The Gap in term of mobility for disabled travellers in France, Paper at the 12th WCTR, July 11–15, Lisbon

Delalex, G. 2004. Living on the motorway. Paper presented at the Alternative Mobility Futures. Conference, Lancaster University, 9–1 January

Delfos, M. 2004. Werken is nepspelen in, *De wereld van het jonge kind*, edited by M. Delfos, 130–33

Denniss, A and Urry, J. 2009. *After the Car*. Cambridge: Polity Press

Department of Infrastructure, Victoria, Australia. 2007. Maintaining mobility: The transition from driver to non-driver. *Policy Framework Report*

Department of Transport UK. 2008a. Carbon pathways analysis: Informing development of a carbon reduction strategy for the transport sector, London

Department for Transport UK. 2010. Transport Trends, 2009 edition and National Travel Survey, London

Department of Transport and Regional Services Australia. 2005. Is the world running out of oil? A review of the debate. *Working Paper 61*

DHV/Smart Agency Company. 2007. Onderzoek Werknemers over Werklocaties

Diekstra, R. and Kroon, M. 2003. Cars and behaviour: Psychological barriers to car restraint and sustainable urban transport, in *The Greening of Urban Transport: Planning for Walking and Cycling in Western* Cities, edited by R. Tolley. Chichester: John Wiley & Sons, 147–57

Dijst, M. and Kwan, M-P. 2002. Accessibility and quality of life: Time-Geographic perspectives, in *Social Dimensions of Sustainable Transport*, edited by K. Donaghy. Aldershot: Ashgate

Dijst, M., de Jong, T. and Ritsema van Eck, J. 2002. Opportunities for transport mode change; an exploration of a disaggregated approach. *Environment and Planning B: Planning and Design*, 29, 413–30

Dinten, van, W. 2006. Met gevoel van realiteit: Over het herkennen van betekenis bij organiseren, Delft: Eburon

Dittrich – Wesbuer, A. und Freudenau, H. 2002. Autofreie Mobilitat – Bedeutung autounabhangiger Wohn- und Lebensformen und Ansatze zur Forderung, in, *Bedeutung psychologischer uns sozialer Einflussfaktoren fur eine nachhaltige Verkehrsentwicklung*, edited by A. Schlaffer. Berlin: Umwelt Bundesamt

Dobbs, L. 2005. Wedded to the car: Women, employment and the importance of private transport. *Transport Policy*, 12, 266–78

Dobbs, L. 2007. Stuck in the slow lane: Reconceptualizing the links between gender, transport and employment. *Gender, Work and Organization*, 4(2), 85–108

Dodson, J. et al. 2003. Transport disadvantage in the Australian metropolis: Towards new concepts and methods. *Working Paper Urban Research Programme*, Brisbane, Griffith University

Dodson, J. and Sipe, N. 2005. Oil vulnerability in the Australian city. *Urban Research Program Paper 6*, Brisbane, Australia: Griffith University

Dodson, J. and Sipe, N. 2006. Shocking the suburbs: Urban location, housing debt and oil vulnerability in the Australian city, *Urban Research Program. Paper 8*, Brisbane, Australia: Griffith University

Dodson, J. and Sipe, N. 2008. Suburban shocks: Assessing locational vulnerability to rising household fuel and mortgage interest costs. Brisbane, Australia: Griffith University

Dunn, J. 1998. Driving forces: the automobile, its enemies, and the politics of mobility, Washington DC, Brookings Institution Press

Dunn, J. 2005. Mobility contested: Ethical challenges for planners, administrators and policy analysts. Paper prepared for the conference on Ethics and Integrity of Governance, Leuven, 2–5 June

Dupuy, G. 1999. *La dependance automobile*, Paris: Anthropos

Dupuy, G. 2011. *Towards Sustainable Transport: The Challenge of Car Dependence.* John Libbey: Eurotext

Durodie, B. 2005. The limitations of risk management. *Tidsskrifte Politik*, 8, 14–21

Durodie, B. 2006. The concept of risk. Risk case studies. Nuffield Trust Global Programme on Health, Foreign Policy and Security

Edensor, T. 2004. Automobility and National Identity; Representation, geography and Driving Practice. *Theory, Culture and Society*, 21(4 /5), 101–20

Elchardus, M. 2004. We lopen een culturele revolutie achter/de symbolische samenleving, Willem Dreeslezing. *Socialisme en Democratie*, 10–21

Elzen, B., Geels, F. and Hofman, P. 2003. Sociotechnische scenarios als hulpmiddel voor transitiebeleid: een illustratie voor het domein van personenmobiliteit, in *Milieubeleid en Technologische Ontwikkeling*, edited by J. Vollebergh et al. Den Haag, Sdu, 269–93

Eurostat 2009. Panorama of Transport, Luxembourg

Ewing, R. et al., 2003. Relationship between Urban Sprawl and Physical Activity, Obesity and Morbidity. *American Journal of Health Promotion*, 18(1), 47–57

Ewing, R. et al. 2007. Growing cooler: Evidence on urban development and climate change. Washington DC, Urban Land Institute

Exel, N. van and Rietveld, P. 2009. Could you also have made this trip by another mode? An investigation of perceived travel possibilities of car and train travellers on the main travel corridors to the city of Amsterdam. *Transportation Research, Part A: Policy and Practice*, 43(4), 374–85

Farrington, J. et al. 1998. *Car Dependence in Rural Scotland*. Edinburgh, Scottish Office, Central Research Unit

Fattouh, B. 2007. The drivers of oil prices: The usefulness and limitations of non-structural model, the demand-supply framework and informal approaches. Oxford: Oxford Institute for Energy Studies, WPM 32

Featherstone, M., Thrift, N. and Urry, J. 2005. *Automobilities*. London: Sage Press

FIA (Foundation for the Automobile and Society). 2004. *Transport and Social Exclusion. A Survey of The Group of Seven Nations.* London

Filarski, R. and Mom, G. 2008. *Van Transport naar Mobiliteit.* Zutphen: Walburg Pers

Filarski, R. 2011. *Shaping Transport Policy: Two centuries of struggle between the public and private sector – A comparative perspective.* Den Haag: Sdu uitgevers

Flamm, M. and Kaufmann, V. 2006. Operationalising the concept of motility: A qualitative study. *Mobilities*, 1(2), 167–89

Flamm, M., Jemelin, C., and Kaufmann, V. 2008. *Travel Behaviour Adaptation Processes during Life Course Transitions.* Lausanne: LaSUr

Fol, S., Dupuy, G., and Coutard, O. 2006. Transport policy and the car divide in the U.K., the U.S. and France: Beyond the environmental debate. *International Journal of Urban and Regional Research*, 31(4), 802–18

Foresight. 2007. Intelligent infrastructure futures – Scenarios toward 2055 – Perspective and Process. London: Office of Science and Technology

Foster, C. 2004. Gendered retailing: A study of customer perceptions of front-line staff in the DIY sector. *International Retail and Distribution Management*, 442–7

Fotel, T. and Thomsen, T. 2004. The surveillance of children's mobility. *Surveillance and Society*, 535–54

Franzen, M. 2004. Retailing in the Swedish city: The move towards the outskirts. *COST Action C 10*, 93–113

Freudendal-Pedersen, M. 2005. Structural stories, mobility and (un)freedom, in *Social Perspectives on Mobility*, edited by T. Thomsen, L. Drewesand, H. Gudmondsson. Aldershot: Ashgate

Freudendal-Pedersen, M. 2008. *Mobility and Daily Life; Between Freedom and Unfreedom*. Aldershot: Ashgate

Freund, P. and Martin, G. 1999. Driving south: The globalization of auto consumption and its social organization in space. *Dept of Sociology, Montclair State University, USA*

Freund, P. and Martin, G. 2004. Walking and motoring: Fitness and the social organisation of motoring. *Sociology of Health and Illness*, 26(3), 273–86

Freund, P. and Martin, G. 2007. Hyperautomobility, the social organization of space and health. *Mobilities*, 2, 37–49

Freund, P. and Martin, G. 2008. Fast cars / Fast foods: Hyperconsumption and its health and environmental consequence. *Social Theory and Health*, 6, 309–22

Fujii, S. and Kitamura, R. 2003. What does a one-month free bus ticket do to habitual drivers? *Transportation*, 30, 81–95

Funk, W. 2009. Mobilitat von Kindern und Jugendlichen. Langfristigre Trends der Andering ihres Verkehrverhaltens, Paper 10 Fachkonferenz; Junge Menschen und Mobilitat, 19/20-11, Munchen

Furedi, F. 2001. *Paranoid Parenting: Abandon your Anxieties and be a Good Parent.* London: Allen Lane

Furedi, F. 2002. *Culture of Fear: Risk Taking and the Morality of Low Expectations.* London: Routledge

Furedi, F. 2004. *Therapy Culture: Cultivating Vulnerability in an Uncertain Age.* London: Routledge

Galbraith, J.K. 1992. *The Culture of Contentment.* Boston: Houghton Mifflin & Co.

Garling, T. and Loukopoulus, P. 2005. Are Car Users Too Lazy to Walk? *TRB Research Record* 1926

Garling, T. and Steg, L. (eds), 2007. *Threats from the Car Traffic to the Quality of Urban Life*. New York: Elsevier

Garreau, J. 1991. *Edge City: Life on the New Frontier*. New York: Doubleday

Gartman, D. 2004. Three ages of the automobile: The cultural logics of the car. *Theory, Culture and Society*, 21(4/5), 169–96

Gatersleben, B. 2007. Affective and symbolic aspects of car use, in *Threats from Car Traffic to the Quality of Urban Life*, edited by T. Garling and L. Steg. New York, Elsevier, 219–33

Geels, F. and Kemp, R. 2006. Dynamics in socio-technical systems: typology of change processes and contrasting case-studies. *Technology and Society*, 441–55

Geels, F. and Schot, J. 2007. Typology of socio-technical transition pathways. *Research Policy*, 36(3) 399–417

Geldof, D. 2008. De financiele crisis en de risicomaatschappij. *Samenleving en Politiek*, 18(9) 30–37

Genre-Grandpierre, C. and Josselin, D. 2008. Dependance a l'automobile, tensions dans les mobilités et strategies des ménages. *Cybergeo Journal of Geography*, 2

Gentili, C. 2003. Transport and social exclusion: A G7 comparison, an overview of the Italian position. Paper presented at the Annual European Transport Conference, 8–10 October, Strasbourg, France

Gershuny, J. 2005. Busyness as the badge of honour for the new superordinate working class. *Social Research*, 72, 287–314

Geurs, K. 2006. Accessibility, land use and transport; Accessibility evaluation of land use and transport developments and policy strategies, thesis, Delft: Eburon

Giddens, A. 1991. *Modernity and Self-Identity; Self and Society in the Late Modern Age*. Palo Alto, CA: Stanford University Press

Giddens, A. 2008a. *The Politics of Climate Change*. Cambridge: Polity Press

Giddens, A. 2008b. The Politics of Climate Change: National responses to the challenge of global warming. London: Policy Network Paper

Giger, M. 2008. Une perspective de genre sur la mobilité quotidienne. Lausanne: *Memoire Université de Lausanne*

Gladwell, M. 2000. *The Tipping Point: How Little Things Make A Big Difference*. New York: Little Brown and Company

Glazebrook, G. and Newman, P. 2008. Coping WITH PEAK OIL AND GLOBAL WARMING in Australian cities. ACSP-AESOP 4th Joint Congress, 6/11-7

Goldberg, H. 2001. State and county supported car ownerships programs can help low-income families secure and keep jobs. Centre on Budget and Policy Priorities

Goodstein, D. 2007. *Out of Gas: The End of the Age of Oil*. New York: W.W. Norton & Co.

Goodwin, P. 1997. Mobility and car dependency, in *Traffic and Transport Psychology, Theory and Application*, edited by T. Rothengatter and E. Vaya. New York: Pergamon Press

Gorham, R. 2002. Car dependence as a social problem: A critical essay on the existing literature and future needs, in *Social Change and Sustainable Transport*, edited by W. Black and P. Nijkamp. Bloomington: Indiana University Press, 107–15

Gorris, T. and Rietveld, P. 2007. Transitie naar duurzame mobiliteit; ruimtelijke ontwikkeling en bereikbaarheid. *Paper voor Colloquim Vervoersplanologisch Speurwerk*

Gorti, R. 2004. An analysis of travel trends of the elderly and zero-vehicle households in the United States, thesis, Tampa: University of South Florida

Götz, K. 1999. Mobilitatsleitbilder und Verkehrsverhalten. *Pro Velo*, 59/1999, 10–16

Götz, K. et al. 2002. Mobility styles in leisure time, Frankfurt, ISEO Paper

Graham, D. and Glaister, D. 2002. The demand for automobile fuel: A survey of elasticities. *Journal of Transport Economics and Policy*, 36, 1–25

Green4Sure. 2007. Het Groene Energieplan. *Achtergrondrapport*

Greene, D. 2008. Future prices and availability of transport fuels. *ITF/OECD: Oil Dependence, Round Table 139*, 115–45

Greene, D., Hopson, J. and Li, J. 2003. Running out of and into oil: Analyzing global oil depletion and transition through 2050. Paper Oak Ridge Laboratory, US

Greene, D., Hopson, J. and Li, J. 2006. Have we run out of oil yet? Oil peaking analysis from an optimist's perspective. *Energy Policy* 34, 515–31

Grieco, M. 2003. Transport and social exclusion: New policy grounds, new policy options. Keynote Paper, 10th International Conference on Travel Behaviour Research. Lucerne, 10–15 August

Grieco, M. and Raje, F. 2004. Stranded mobility and the marginalisation of low-income communities. Paper presented at the conference on Urban Vulnerability and Network Failure, University of Salford

Hagerstrand, T. 1970. What about people in regional science? *Papers and Proceedings of the Regional Science Association*, 24, 7–21

Hagman, O. 2001. Mobilising meanings of mobility: Car users constructions of the goods and bads of car use. *ESA Conference New Technologies and New Visions*, Helsinki

Hagman, O. 2004. Alternative mobilities in the networking of everyday life. Conference Alternative Mobility Futures, Lancaster University, 9-11/1

Halleux, J.M., Bruck, L. and Mairy, N. 2002. La periurbanisation residentielle en Belgique a la lumiere des contexts Suisse et Danois: enracinement, dynamiques centrifuges et regulations collective. *Belgio*, 4, 333–56

Hamers, D. and Nabielek, K. 2006. Bloeiende Bermen: Verstedelijking langs de snelweg, Ruimtelijk Planbureau. Den Haag: NAi uitgevers

Handy, S. 2005. Smart growth and the transportation-land use connection: What does the research tell us? *International Regional Science Review*, 28(2), 146–67

Handy, S., Weston, L. and Mokhtarian, P. 2005. Driving by choice or necessity: The case of the soccer mom and other stories. *Transportation Research, Part A*, 34, 183–204

Hardin, G. 1968. The tragedy of the commons. *Science* 162, 1243–8

Harms, L. 2003. Mobiel in de tijd. Op weg naar een auto-afhankelijke samenleving. Den Haag: SCP

Harms, L. 2006a. Anders Onderweg ? Mobiliteit van allochtonen en autochtonen vergeleken, Den Haag: SCP

Harms, L. 2006b. Op weg in de vrije tijd; context, kenmerken en dynamiek van vrijetijdsmobiliteit, Den Haag: SCP

Harms, L. 2008. Overwegend Onderweg: de leefsituatie en de mobiliteit van Nederlanders. Den Haag: SCP

Harms, S. 2003. From routine choice to rational decision making between mobility alternatives. Conference Paper, Swiss Transport Research Conference, Ascona, March

Harvey, D. 1989. *The Condition of Post Modernity*. Oxford: Blackwell

Healey, P. 2007. *Urban Complexity and Spatial Strategies: Towards a Relational Planning for Our Times.* New York: Routledge

Heijmans, T. 2007. *La Vie Vinex.* Amsterdam: L.J. Veen uitgevers

Heine, H. and Mantz, R. 1999. Die Mutter und das Auto, PKW-Nutzung im Kontext geslechtsspezifischen Arbeitsteilung, Gottingen, *SOFI Mitteilungen 27*

Heine, H. and Mantz, R. 2000. Mobilitat und Grenze des Autoverzichts. Abschlussbericht Universitat Gottingen

Heine, H., Mantz, R. and Rosenbaum, W. 2001. Mobilitat im Alltag. Warum wir nicht vom Auto lassen, Frankfurt/New York

Heinze, G. 2007. Offenticher Verkehr und demographischer Wandel: Chancen fur Nordostdeutschland, in *Die Zukunft der Infrastrukturen in landlichen Raumen*, edited by S. Beets. Berlin: Berlin-Brandenburgischen Akademie der Wissenschaften, S 21-30

Hess, C. 2008. Mapping the commons. Paper presented at the 12th Biennial Conference of the International Association for the Study of the Commons. Cheltenham, 12–14 July

Heywood, J. 2008. More sustainable transportation: The role of energy efficient vehicle technologies. Report prepared for ITF/OECD Leipzig 28/30-5

Heywood, J. and Bandivadekar, A. 2005. New vehicles; How soon can they make a difference? *Energy and Environment: Journal of MIT Laboratory for Energy and the Environment*, 3, 1–2

Hickman, R. and Banister, D. 2005. Towards a 60% reduction in UK Transport Carbon Dioxide Emissions: A scenario-building and backcasting approach, paper from VIBAT study.

Hickman, R. and Banister, D. 2005. Looking over the horizon: Transport and reduced CO_2 emissions in the U.K. by 2030. Paper on ETC Conference

Hickmann, R. and Schwanen, T. 2010. Enabling sustainable mobilities: social, cultural and experiental dimensions, and the role of planning. Paper for 12th WCTR, July 12–15, Lisbon

Hilbers, H., Snellen, D. and Hendriks, A. 2006. Files en de ruimtelijke inrichting van Nederland, Ruimtelijk Planbureau. Den Haag: Nai

Hine, J. and Mitchell, F. 2001. Better for everyone? Travel experiences and transport exclusion. *Urban Studies*, 3(2) 319–32

Hine, J. and Mitchell, F. 2003. *Transport Disadvantage and Social Exclusion.* Aldershot: Ashgate

Hirsch, R. 2007. Peaking of world oil production: recent forecasts. *World Oil*, Vl.228, (4), 113–20

Hirsch, R., Bezdek, R. and Wendling, R. 2005. Peaking of World Oil Production; Impacts, Mitigation and Risk Management, Morgantown, WV, U.S. Department of Energy, National Energy Technology Laboratory

Hjortol, R. 2005. Mobility in Daily Life, the Car and Use of ICT for Family Logistics. Paper submitted to the 45th Congress of The European Regional Science Association, Amsterdam, August

Hjortol, R. and Fyhri, A. 2009. Do organized leisure activities for children encourage car use? *Transportation Research, Part A, Policy and Practice*, 43(2), 209–18

Holden, E. 2004. Towards sustainable consumption; do green households have smaller ecological footprints? *International Journal of Sustainable Development*, 7(1), 44–58

Holden, E. 2007. *Achieving Sustainable Transport*. Aldershot: Ashgate

Hortulanus, R. and Machielse, J. 2001. Op het snijvlak van de fysieke en de sociale leefomgeving, Den Haag

Hughes, J., Knittel, C. and Sperling, D. 2006. Evidence of a shift in the short-run price elasticity of gasoline demand, *National Bureau of Economic Research, Working Paper*, Cambridge, MA.

Huttenmoser, M. 2005. Der Tanz mit dem Bandel, oder Wass heist Kindgerechte Sicherheitspolitik. *Vortrag zur Grundung des Netzwerks 'Kind und Verkehr'*.

IEA 2007a. Energy use in the new millennium: Trends in IEA countries, Paris

IEA 2007b. World energy outlook, Paris

IEA 2008a. Worldwide trends in energy use and efficiency: Key insights from IEA Indicator Analysis, Paris

IEA 2008b. Halving the costs of halving emissions, considering price gaps in climate policy, September, Paper in support of the G8 Plan of Action, Paris

IFMO 2007. Mobilitat 2025: Der Einfluss von Einkommen, Mobilitätskosten und Demografie, Berlin

Imanashi, Y. 2003. Transport and social exclusion. The Japanese experience, FIA Foundation

Infas und DLR 2010. Germany: Mobilitat im Deutschland. Ergebnisbericht; Struktur-Aufkommen-Emissionen-Trends. Berlin-Adlershof

Innes, J. and Booher, D. 2003. The impact of collaborative planning on governance capacity, IURD Working Paper, Berkeley, University of California

Innes, J. and Booher, D. 2010. *Planning With Complexity. An Introduction to Collaborative Rationality for Public Policy*, New York: Routledge

ITF/OECD 2008a. Oil dependence. Is transport running out of affordable fuel? *Round Table 139*, Paris

ITF/OECD 2008b. Biofuels: Linking support to performance, *Round Table 138*, Paris

ITF/OECD 2008c. Transport and energy: The challenge of climate change. *Research Findings, delivered at the ITF Symposium* 28/30 – 5, Leipzig

ITF 2009. Trends in the transport sector, Paris, 1970–2009

ITS 2003. Verknocht aan de auto ? Den Haag: Adviesdienst Verkeer en Vervoer

Izumiyana, H., Ohmori, N. and Harata, N. 2007. Space-time accessibility measures for evaluating mobility-related social exclusion of the elderly. Paper, Hiroshi Institute, University of Tokyo

Jackson, T. 2004a. Motivating Sustainable Consumption: a review of evidence on consumer behaviour and behavioural change. Paper, Centre for Environmental Strategy, University of Surrey

Jackson, T. 2004b. Consuming Paradise? Unsustainable consumption in cultural and social-psychological context, in *Driving Forces of and Barriers to Sustainable Consumption*, edited by K. Hubacek et al. Proceedings of an International Workshop, Leeds, 5–6 March

Jackson, T. 2004c. Models of Mammon; A cross-disciplinary survey in pursuit of the 'Sustainable Consumer'. *Working Paper Series no 2004/1, Centre for Environmental Strategy, University of Surrey*

Jackson, T. 2006. *Readings in Sustainable Consumption. The Earthscan Reader in Sustainable Consumption.* London: Earthscan

Jarvis, H. 2004. City time: managing the infrastructure of everyday life. ESRC Work Life and Time in the New Economy, seminar paper

Jarvis, H. 2005. Moving to London: Time household coordination and the infrastructure of everyday life. *Time and Society*, 14(1), 133–54

Jeekel, H. 2011. De autoafhankelijke samenleving, Delft, Eburon

Jesse, J.H. and vd Linde, C. 2008. Oil turbulence in the next decade. An essay on high oil prices in a supply-constrained world. *Clingendael International Energy Programme*

Johnson, V. 2007. Car ownership and social exclusion in Australia. Paper for CAITR conference 5-8/ 12

Karsten, L. and De Stigter-Speksneijder, M. 2006. Vinexwijken: de professionele kritiek en de dagelijkse woonpraktijk, een essay, Utrecht, NETHUR

Karsten, L., Kuiper, W. and Reubsaet, F. 2001. Van de Straat: de relatie jeugd en openbare ruimte verkend. Assen: Van Gorcum

Karsten, L. and Van Vliet, W. 2006. Increasing children's freedom of movement. *Children, Youths and Environments*, 16(1), 69–72

Kaufman, V. 2002. *Re-thinking Mobility*. Aldershot: Ashgate Publishing

Kaufmann, V. and Flamm, M. 2002. Famille, temps et mobilité : état de l'art et tour d'horizon des innovations. *Recherche realisée a l'intention de la CNAF et de l'institut pour la Ville en Mouvement*

Kaufmann, V. et al. 2004. Mobilité et motilité: de l'intention a l'action. *Cahier LASUR*, June

Kaufmann, V., Bergman, M. and Joye, D. 2004. Motility: Mobility as capital. *International Journal of Urban and Regional Research*, 28(4), 745–56

Kawabata, M. and Shen, Q. 2005. Job accessibility as an indicator of auto-oriented urban structure: a comparison of Boston and Los Angeles with Tokyo. *Environment and Planning B*, 33, 115–30

Kemming, H. and Borbach, C. 2003. Transport and social exclusion: A G-7 comparison. An overview of the German position, FIA Foundation.

Kennisinstituut voor Mobiliteitsbeleid 2007. Mobiliteitsbalans 2007, Den Haag

Kennisinstituut voor Mobiliteitsbeleid 2008a. Olieprijzen, economische groei en mobiliteit, Den Haag

Kennisinstituut voor Mobiliteitsbeleid 2008b. Mobiliteitsbalans 2008 ; congestie in perspectief, Den Haag

Kennisinstituut voor Mobiliteitsbeleid 2008c. Verkenning Autoverkeer 2012, Den Haag

Kennisinstituut voor Mobiliteitsbeleid 2008d. Vrijetijdsverkeer in perspectief. De relatieve economische waarde van het vrijetijdsverkeer, Den Haag

Kennisinstituut voor Mobiliteitsbeleid 2008e. Grijs op reis; over de mobiliteit van ouderen, Den Haag

Kennisinstituut voor Mobiliteitsbeleid 2008f. Blijvend anders onderweg, Den Haag

Kennisinstituut voor Mobiliteitsbeleid 2009a. Mobiliteitsbalans 2009, Den Haag

Kennisinstituut voor Mobiliteitsbeleid 2009b. Het imago van het openbaar vervoer, Den Haag

Kennisinstituut voor Mobiliteitsbeleid 2009c. Bij het scheiden van de markt; vraagontwikkelingen in het personen- en goederenvervoer, Den Haag

Kennisinstituut voor Mobiliteitsbeleid 2009d. Het belang van het openbaar vervoer ; de maatschappelijk effecten op een rij, Den Haag

Kennisplatform Verkeer en Vervoer 2005. Openbaar Vervoer en Doelgroepenvervoer.

Kennisplatform Verkeer en Vervoer 2007. Ontwikkeling van het aanbod en gebruik van OV – diensten vanaf 2000 tot 2006

Kim, A. 2002. Taken for a ride: Subprime lenders, automobility, and the working poor, Progressive Policy Institute

Klinger, T., Kenworthy, J. and Lanzendorf, M. 2010. Mobility culture in urban areas – a comparative analysis of German cities. WTRC Paper, 11–15 July, Lisbon

Klöckner, C. 2005. Das Zusammenspiel von Gewohnheiten und Normen in der Verkehrmittelwahl – ein integriertes Norm – Aktivations Modell und Seine Implikationen fur Interventionen, Bochum, Dissertation Ruhr Universität

Knulst, W. 1984. Tijdsbesteding en Overheidsbeleid, Den Haag: SCP rapport

Kohler, J. 2006. Transport and the environment: Policy and Eeconomic considerations in UK. Foresight Intelligent Infrastructure Systems Project

Korsu, E. and Massot, H. 2006. Rapprocher les ménages de leurs lieux de travail: les enjeux pour la regulation de l'usage de la voiture en France. *Les Cahiers Scientifiques du Transport*, 61–90

Kotkin, J. 2010. The war against suburbia. *The American*, 21 January

Kromer, M. and Heywood, J. 2007. Electric power trains: Opportunities and challenges in the U.S. light duty fleet. Cambridge, MA: MIT

Kurani, T. and Turrentine, T. 2002. Marketing clean and efficient vehicles: A review of social marketing and social science approaches. Paper, Institute of Transportation Studies, University of California, Davis

Kurani, T. and Turrentine, T. 2004. Automobile buyers' decisions about fuel economy and fuel efficiency. Paper, Institute of Transport Studies, University of California, Davis

Laermans, R. 2006. Stedelijkheid in de veralgemeende moderniteit, *Paper Centrum voor Cultuursociologie*, K.U. Leuven

Laessoe, J. 2001. The need for mobility, in *Mobility and Transport: An Anthology*, edited by H. Drewes-Nielsen and H. Oldrop. Copenhagen: The Danish Transport Council, blz. 99–106

Laherrere, J. 2004. Perspectives energetiques et scientifiques. *Presentation au club des jeunes dirigeants de Quimper*, 22–4

Lang, R. 2003. *Edgeless Cities: Exploring the Elusive Metropolis*. Washington DC: Brookings Institution

Lareau, A. 2002. Invisible inequality: Social class and childrearing in black families and white families. *American Sociological Review*, 67(5) 747–6

Larsen, J., Urry, J. and Axhausen, K. 2008. Coordinating face-to-face meeting in mobile network societies. *Information, Communication & Society*, 11(5), 640–58

Larsen K. and Gilliland, J. 2008. Mapping the evolution of 'food deserts' in a Canadian city: Supermarket accessibility in London, Ontario, 1961–2005. *International Journal of Health Geographics*, 7(1), 1436 e.v

Laurier, E. 2004. Doing office work on the motorway. *Theory, Culture and Society*, 21, 261–77

Laurier, E. 2005. Habitable cars: What we do there. Lecture given at the 5th Social Studies of Information Technology, London School of Economics, April

Leggett, J. 2006. What they don't want you to know about the coming oil crisis. *Independent*, 20 January

Levelt, P. 2001. Boze agressie in het verkeer. Een emotietheoretische benadering. *Justitiële Verkenningen*, 27, 95–108

Levelt, P. 2002. Literatuurstudie naar emoties in het verkeer, SWOV, R-2002-31

Levy, D. and Kolk, A. 2002. Strategic responses to global climate change: Conflicting pressures on multinationals in the oil industry. *Business and Politics*, 4(3), 275–300

Levy, D. and Rothenberg, S. 2007. Heterogeneity and change in environmental strategy: Technological and political responses to climate change in the global automobile industry, in *Beyond Isomorphism, Institutional and Strategic Perspectives*, edited by A. Hofman and M. Ventresa. Palo Alto, CA:, Stanford University Press, Chapter 7

Limbourg, M. 2005. Mobilitat im Kindesalter. Rapport, Universität von Duisburg

Linde, M. 2007. Energie: de eeuw van mijn moeder, in *IJsberenplaag op de Veluwe*, Essays over de Toekomst. COS Den Haag

Ling, R. and Haddon, L. 2001. Mobile telephony, mobility and the coordination of everyday life. Paper presented at Rutgers University, 18/19-4

Lipovetsky, G. and Charles, S. 2004. *Les Temps Hypermodernes*, Paris: Grasset

Litman, T. 2002. *Social Inclusion as a Planning Goal in Canada*. Victoria (Canada): Victoria Transport Research Institute

Litman, T. 2008. Appropriate Response to rising Fuel Prices. Victoria (Canada), Victoria Transport Policy Institute

Locke, J. 1998. *The De-voicing of Society: Why Don't we Talk to Each Other Anymore?* New York: Simon & Schuster

Lomansky, L. 1995; Autonomy and automobility. *The Independent Review*, II, 1, 5–28

Loorbach, D. 2008. Why and how transition management emerges. Paper Conference Socio-Ecological Research, Berlin

Loorbach, D. 2010. Transition management for sustainable development, prescriptive complexity-based governance framework. *Governance*, 161–83

Loorbach, D. and Rotmans, J. 2006 Managing transitions for sustainable development, in *Understanding Industrial Transformation: Views from Different Disciplines*, edited by X. Olsthoorn et al. Dordrecht: Springer

Lopez-Ruiz, H. and Crozet, Y. 2010. Sustainable transport in France: is a 75 % reduction in CO_2 emissions possible? *Transportation Research Record*, 2163, 124–32

Lucas, K. 2003. Transport and social exclusion: A G-7 comparison. An overview of the UK position. FIA Foundation

Lucas, K. 2008. Scoping study of actual and perceived car dependence and the likely implications for livelihoods, lifestyles and well-being of enforced reductions in car use. Paper TRB 2009

Lucas, K. and Jones, P. 2009. *The Car in British Society*. RAC Foundation.

Lucas, K., Grosvenor, T. and Simpson, R. 2001. *Transport, the Environment and Social Exclusion*. London: Joseph Rowntree Foundation

Lucas, K. and LeVine, S. 2009. The car in British society. RAC Foundation, Working paper 2: Literature Review

Lucas, K., Blumenberg, E. and Weinberger, R. (eds). 2011. *Auto Motives: Understanding Car Users Behaviour*. Bingley, UK: Emerald

Luijendijk, J. 2010. Artikelenserie over de elektrische auto, in NRC Handelsblad, January–February

Lupi, T. 2005. Community Light: territorial ties and local participation, Paper prepared for the OTB conference Doing, Thinking, Feeling Home: The Mental Geography of Residential Environments, Delft 14/15-10

Lupi, T. and Musterd, S. 2004. The Suburban Community Question. Paper prepared for the ESF conference Cohesive Neighbourhoods and Connected Citizens, in European Studies, Bristol 17/18-6

Lutsey, N. and Sperling, D. 2009. Greenhouse gas mitigation supply curve for the United States for transport versus other sectors. *Transportation Research, Part D*, 14, 222–9

Lynch, M. 2004. The new pessimism about petroleum resources: Debunking the Hubbert Model (and Hubbert Modellers). *Minerals and Energy*, 18(1), 21–32

Lyons, G. 2003a. Future mobility – it's about time. Paper presented at the Universities Transport Study Group Conference, Loughborough, January 2003

Lyons, G. 2003b. Transport and Society, inaugural lecture, 1 May, Bristol, University of the West of England

Lyons, G. 2003c. The introduction of social exclusion into the field of travel behaviour. *Transport Policy*, 339–42

Lyons, G. and Urry, J. 2006. Foresight; the place of social science in examining the future of transport. Paper based at Evidence-based Policies and Indicator Systems, 11–13 July, London

Lyons, G. et al. 2008. Public Attitudes to transport; knowledge review of existing evidence. Centre of Transport & Society, Bristol, University of the West of England

Mackett, R. et al. 2002. Understanding the car dependency impacts of children's car use. Paper written for the workshop on Children and Traffic, Copenhagen, 2–3 May

Malone, K. 2002. Street Life: Youth, culture and competing uses of public space. *Environment and Urbanisation*, 14(2), 157–68

Manderscheid, K. 2012. Automobile Subjekte. Dortmunder Konferenz Raum- und Forschungsplanung, Technische Universitat Dortmund, 9-10/2

Marletto, J. 2010. Structure, agency and change in the car regime. A review of the literature. Paper, Centro di Recerca Interdipartmentale di Economia della Institutione (CREI), University of Sassari

Martens, K. 2000. Debatteren over Mobiliteit. Over de rationaliteit van het ruimtelijk mobiliteitsbeleid, proefschrift, Nijmegen

Martens, K. 2006. Basing transport planning on principles of social justice. *Berkeley Planning Journal*, 19, 1–17

Maxwell, S. 2001. Negotiations of car use in everyday life, in *Car Cultures*, edited by D. Miller. Oxford: Berg Publishers, 1–35

McDonald, N. 2005. Children's Travel; Patterns and Influences. Thesis, Berkeley, CA: University of California

McMillan, T. et al. 2006. Johnny walks to school – does Jane ? Sex differences in children's active travel to school. *Children, Youth and Environments*, 16, 75–89

Meiklejohn, D. 2008. Addressing Oil Vulnerability through Travel Behaviour Change. Paper presented at the 29th Australasian Transport Research Forum

Meisonnier, J. 2008. Pour faire face a l'automobilité: des transport en commun plus ludiques? in *Automobilité et altermobilités: quels changements?*, edited by F. Clochard, A. Rocci and S. Vincent, Dossiers Sciences Humaines et Sociales, Editions de l'Harmattan 41–18

Merriman, P. 2004. Driving places: Marc Augé, non-places and the geographies of England's M1 motorway. *Theory, Culture and Society*, 21, 245–67

Mesken, J. 2006. Determinants and consequences of drivers' emotions. Leidschendam, SWOV

Messerli, P. and Trosch, M. 2002. Why it is not easy to change mobility behaviour in winter sports? *Revue de Geographie Alpine*, 90(1), 67–82

Mignot, D. 2004. Transport et Justice Sociale. *Reflets et Perspectives de la Vie Economique*, Numero Special Transport et Mobilité, Tome XLIII, 123–31

Miller, H. 2004. Travel chances and social exclusion, Paper, University of Utah

Miller. D. ed. 2001. *Car Cultures*. Oxford: Berg Publishers

Ministerie van Economische Zaken 2008. Energierapport 2008, Den Haag

Ministerie Verkeer en Waterstaat 1990. Tweede Structuurschema Verkeer en Vervoer, Staatsuitgeverij, Den Haag

Ministerie van Verkeer en Waterstaat 2009. Plan van Aanpak Elektrisch Rijden, Den Haag

Moeckli, J. and Lee, J. 2005. The making of driving cultures, in AAA Foundation of Traffic Safety, Improving Traffic Safety Culture in the USA, Washington, 59–76

Moens, M. 2004. Handelen onder druk; tijd en tijdsdruk in Vlaanderen. *Tijdschrift voor Sociologie* 25(4), 383–416

Moisio, R. 2007. Men in no-man's land: Proving manhood through compensatory consumption, thesis, Lincoln: University of Nebraska

Mokhtarian, P. and Salomon, I. 1999. Travel for the fun of it. *Access*, 15, Fall

Mol, J., de, Lavrysen, L. and Vlassenroot, S. 2006. Toenemend vermogen, topsnelheid en gewicht van auto's en de productaansprakelijkheid van de autoconstructeurs. Paper Colloquim Vervoersplanologisch Speurwerk, Amsterdam, 23–24 November

Molnar, H. 2005. Mobiliteit van ouders met jonge kinderen. *CBS Economische trends*, 4 de kwartaal, 49–51

Mommaas, H. 2003. Vrije tijd in een tijdperk van overvloed, inaugurale rede, Tilburg, Universiteit van Tilburg

Mobiliteitsonderzoek Nederland (MON), 2007, 2008, 2009. AVV/DVS Rijkswaterstaat Rotterdam, Delft

Moriarty, P. and Honnery, D. 2009. Australian car travel: An uncertain future. Paper, 30th Australasian Transport Research Forum

Morris, K. 2006. Research into travel horizons and its subsequent influence on accessibility planning and demand responsive transport strategies in Greater Manchester. Paper, for ETC Strasbourg

Motivaction 2002. Mobiliteitsbeleving en mentaliteitsprofielen van de Nederlandse bevolking, Amsterdam

Motte-Baumvol, B. 2007. La dependance automobile pour l'acces aux services aux ménages en grande couronne francilienne. Paris, these, Université de Paris I

Mourik, R. et al. 2005. E-magining future energy infrastructures in the Dutch Transport sector: An analysis of stakeholder perspectives on future energy infrastructures in the transport sector, Petten, ECN-C-05-051

Mulder, C. and Kalmijn, T. 2004. Even bij oma langs; NKPS laat zien hoe ver familieleden van elkaar wonen. *Demos, bulletin over bevolking en samenleving*, 20, 78–80

Nationale Franchisegids 2007. Onderdeel Formido, interview met Dick Schreuders, Manager expansie en franchise.

National Renewable Energy Laboratory 2003. Consumer views on Transportation and Energy. Golden, CO., USA

Newman, P. and Kennworthy, J. 1989. *Cities and Automobile Dependence: An International Sourcebook*. Aldershot: Gower

Newman, P. and Kenworthy, J. 1999. *Sustainability and Cities: Overcoming Automobile Dependence.* Washington, DC: Island Press

Nieuwenhuis, P. no date. Is sustainable car making possible? Paper at the 10th International Conference of the Greening of Industry Network, June 23–26, Goteborg, Sweden

Nio, I. 2006. Vinex: tussen identificatie en onthechting. *Ruimte in Debat, Ruimtelijk Planbureau,* 5

Nivola, P. 2009. The Long and Winding Road: Automotive Fuel Economy and American Politics. New York: Governance Studies at Brookings Institution

Noack, E. 2010. Stuck in the countryside? Women's transport mobility in rural Aberdeenshire, Scotland – experiences, behaviour and needs. *Jahrbuch der Osterreichischen Gesellschaft fur Agrarokonomie,* Band 19

Nooteboom, S. 2006. Adaptive networks: The governance for development, Proefschrift, Delft, Eburon

Nooteboom, S. and Teisman, G. 2007. Bestuurlijk Vermogen. De cruciale betekenis van adaptieve netwerken, Paper, Erasmus Universiteit, Rotterdam

Norland, I., Holden, E., Lafferty, W. 2004. Consumption of energy and transport in urban households: The role of urban planning v. 'green consumerism' in promoting sustainable consumption, in *Innovation in Life Cycle Engineering and Sustainable Development,* edited by E. Hartwich and G. Peters. Oslo, 139–60

Normark D. 2006. Tending to mobility – intensities of staying at the petrol station. *Environment and Planning A,* 38, 241–52

Nuvolati, G. 2003. Resident and non-resident populations: Quality of life, mobility and time policies. *Journal for Regional Analysis and Policy,* 33(2), 67–83

O'Brien, C. and Gilbert, R. 2003. Kids on the move in Hutton and Peel. Toronto, Centre for Sustainable Transportation, The Ontario Trillenium Foundation

Oenen, van G. 2007. Interactive metal fatigue: The interpassive transformation of modern life. Presentation at studiedag Theaterwetenschap, 31 August

Office of Science and Technology. Foresight, Mobility Programme, 2010

Ohnmacht, T. et al. 2008. Mobility styles in leisure time – Target groups for measures towards sustainable leisure travel in Swiss agglomerations. Conference paper, Swiss Transport Research Conference, October 15–17

Ohnmacht, T., Maksim, H. and Bergman, M. 2009. *Mobilities and Inequality.* Aldershot: Ashgate

Ohnmacht, T. 2009. Mobilitatsbiografie und netzwerkgeografie: Kontaktmobilitat in ego-zentrienten netzwerken, Dissertation, Zurich

Olden, H. 2010. Uit voorraad leverbaar: De overgewaardeerde rol van bouwrijpe grond als vestigingsplaatsfactor bij de planning van bedrijventerreinen, proefschrift, Utrecht, Geomedia

Orfeuil, J.P. 2004a. Transports, Pauvretés, Exclusions; Pouvoir Bouger pou s'en sortir, La Tour d'Aigues, Editions de l'Aube

Orfeuil, J.P. 2004b. Mobility, poverty and exclusion in France. FIA Foundation

Orfeuil, J.P. 2010. La Mobilité, nouvelle question sociale?, in *Sociologes* (online), accessed, 6 June 2011

Orsato, R. and Wells, P. 2007. U-turn: the rise and demise of the automobile industry. *Journal of Cleaner Production*, 15, blz 994–1006

Ortar, N. 2008. Entre ville et campagne, le difficile equilibre des periurbaines lointaines. *Metropoles* (online), accessed 6 June 2011

Ory, D. and Mokhtarian, P. 2005. When is getting there half the fun? Modelling the liking for travel. *Transportation Research, Part A*, 39, 97–123

Ostrom, E. 1999. Coping with the tragedies of the Commons. *Annual Review of Political Science*, 2, 493–535

Ostrom, E. 2003. How types of goods and property rights jointly affect collective action. *Journal of Theoretical Politics* 15(3), 239–70

Packer, J. 2008. *Mobility without Mayhem: Safety, Cars and Citizenship*. Durham, NC: Duke University Press

Paterson, M. 2006. *Automobile Politics*. Cambridge: Cambridge University Press

Pel, B. 2009a. The complexity of self-organization: boundary judgments in traffic management, in *Managing Complex Governance Systems*, edited by G. Teisman, A. van Buuren, and L. Gerrits. New York: Routledge, 116–34

Pel, B. and Teisman, G. 2009b. Governance of transitions as selective connectivity, Paper IRPSM Conference

Pel, B. and Teisman, G. 2009c. Mobiliteitsbeleid als klimaatbeleid of watermanagement; zelforganisatie als aangrijpingspunt voor effectieve beleidsmatige interventies, Paper, Colloquim Vervoersplanologische Werkdagen, Antwerpen, 19/20-11

Pernack, R. 2005. Offentlicher Raum und Verkehr; Eine sozialtheoretische Annaherung, Berlin, *WZB discussion paper*, SP III 2005-104

Peters, P. 2006. *Time, Innovation and Mobilities*. London: Routledge

Petersen, L. and Andersen, A. 2009. Socio-cultural barriers to the development of a sustainable energy system – the case of hydrogen. Research Note National Environmental Research Institute, Aarhus, Denmark

Planbureau voor de Leefomgeving. 2009. Elektrisch Rijden: een evaluatie van systeemopties, Den Haag

Ploeg, van der H. et al. 2008. Trends in Australian children travelling to school, 1971–2003: burning petrol or carbohydrates ? *Preventive Medicine*, blz. 60–62

Pooley, C., Turnbull, T. and Adams, G. 2005. *Mobility in Everyday Life: Changes in Everyday Mobility in Britain in the 20th Century*. Aldershot: Ashgate

Polzin, S. 2006. The case for moderate growth in vehicle miles of travel: A critical juncture in U.S. travel behavior. Tampa, University of South Florida

Putnam, H. 2000. *Bowling Alone: The Collapse and Revival of American Community*. New York: Simon & Schuster

Raje, F. 2003. *Negotiating the Transport System: Users Contexts, Experiences and Needs*. Aldershot: Ashgate

Raje, F., Grieco, M., Hine, J., Preston, W. 2004. *Transport, Demand Management and Social Exclusion*. Aldershot: Ashgate

Rammler, S. 2003. So unvermeidlich wie die Kautzchen in Athen: Anmerkungen zur Soziologie des Automobils. *IVP – Schriften*, Berlin, December

Rammler, S. 2008. Die wahlverwandschaft of modernity and mobility, in *Tracing Mobilities*, edited by W. Canzler, V. Kaufman and S. Kesselring. Aldershot: Ashgate, 57–77

Rammler, S. 2010. Reinventing mobility: 14 Theses on mobility policy. Paper, Institut fur Transportation Design, Braunschweig

Redshaw, S. 2004. Roads for change: Changing the car and its expressions. Paper, presented to the Social Change in the 21st Century Conference, Queensland University of Technology, October

Redshaw, S. 2008. *In the Company of Cars: Driving as a Social and Cultural Practice*. Aldershot: Ashgate

Reijndorp, A. et al. 1998. *Buitenwijk: Stedelijkheid op afstand*. Rotterdam: Nai

Remoy, H. 2010. Out of office: A study on the cause of office vacancy and transformation as a means to cope and prevent, thesis, Amsterdam University Press

Richardson, A. and Empt, E. 1997. Car availability: Accounting for temporal variations, 25th PTRC European Transport Forum, Brunel University, England

Ridgewell, C., Sipe, N. and Buchanan, N. 2005. School travel modes in Brisbane, Urban Research Programme, Brisbane, Australia, Griffith University, Research Paper 4

Rittel, H. and Webber, M. 1973. Dilemmas in a general theory of planning. *Policy Sciences*, 4, 155–69

RIVM (Geurs, K. and van Wee, G.P.) 2000. Environmentally sustainable transport; implementation and impacts for the Netherlands for 2030, Bilthoven

Roberto, E. 2008. Commuting to opportunity: The working poor and commuting in the U.S., New York: Brookings Institution

Robinson, B. and Mayo, S. 2008. Peak oil and Australia: Probable impacts and possible options. Paper, Australian Association for the Study of Peak Oil & Gas

Rosa, H. 2005. *Beschleunigung, Die Veranderung der Zeitstrukturen in der Moderne*. Frankfurt am Main: Suhrkamp Verlag

Rosenbloom, S. and Stahl, A. 2002. Automobility among the elderly: The convergence of environmental, safety, mobility and community design issues. *EJTIR*, 2(3–4), 197–213

Rougé, L. and Bonnin, S. 2009. Les 'captifs' du periurbain 10 ans après. Rapport CERTU 08-26, Lyon

Ruimtelijk Planbureau. 2005. Winkelen in Megaland. Den Haag: NAi

Runge, D. 2005. Mobilitatsarmut in Deutschland? Schrifte des Fachgebietes Integrierte Verkehrsplanung.

RWS-AVV 2006. Vervoerswijzekeuze op ritten tot 7,5 kilometer, Rotterdam, Argumentaties van autobezitters voor de keuze van de auto cq. de fiets bij het maken van een korte rit, Rotterdam

Sager, T. 2005. Footloose and Forecast-free: hypermobility and the planning of society. *European Journal of Spatial Development*, September, 465–88

Salon, D. and Sperling, D. 2008. City carbon budgets: A policy mechanism to reduce vehicle travel and greenhouse gas emissions. Prepared for the ITF Symposium Leipzig, May

Sanchez, T., Stolz, R. and Ma, J. 2003. Moving to equity: Addressing inequitable effects of transportation policies on minorities. Cambridge, MA: The Civil Rights Project at Harvard University.

Sanchez, T. and Brenman, M. 2007. Transportation equity and environmental justice: Lessons from hurricane Katrina. Paper presented at The State of Environmental Justice in America 2007 Conference, Washington DC, March 29–31

Sandalow, D. 2008. Ending oil dependence: Protecting national security, the environment and the economy. Washington: The Brookings Institution

Sandqvist, K. and Kristrom, S. 2001. *Getting Along Without A Family Car. The Role of an Automobile in Adolescents Experiences and Attitudes*. Inner city Stockholm. KFB – Kommunukationforskningsberedningen, Stockholm.

Sandqvist, K. 2002. How does a family car matter? Leisure, travel and attitudes of adolescents in inner city Stockholm. *World Transport Policy and Practice*, 8(1), 11–18

Sanne, C. 2005. The consumption of our discontent. *Business Strategy and the Environment*, 14(5), 315–23

Sarmento, J. 2006. Driving through car geographies. *Aurora Geographical Journal*, 84–107

Sassen, S. 2002. Introduction: locating cities on global circuits, in *Global Networks, Linked Cities*, edited by S. Sassen. New York and London: Routledge, 1–38

Scheiner, J. 2006. Does the car make elderly people happy and mobile? Settlement structure, car availability and leisure mobility of the elderly. *European Journal of Transport and Infrastructure Research*, 6(2), 151–72

Schipper, L. and Ng, W. 2004. Rapid Motorization in China: Environmental and social challenges, Paper World Resources Institute, Washington

Schokker, T. and Peters, P. 2006. *Hypermobielen, bijdrage aan het Colloquim Vervoersplanologisch Speurwerk*

Scholl, L. 2002. Transportation affordability for low-income populations. Public Policy Institute of California, San Francisco

Schulze, G. 2005. *Die Erlebnisgesellschaft: Kultursoziologie der Gegenwart*. Frankfurt am Main: Campus

Schwanen. T. 2004. How time-pressured individuals cope with disturbing events affecting everyday activities: a literature review and conceptual analysis. Paper, Colloquium Vervoerplanologisch Speurwerk, 25-26-9, Zeist

Schwanen, T. 2007a. Matter(s) of interest: artefacts, spacing and timing. *Geografiska Annaler*, 89B (1), 9–22

Schwanen, T. 2007 b. Gender differences in chauffeuring children among dual-earner families. *The Professional Geographer*, 59(4), 447–62

Schwanen, T. 2008. Managing uncertain arrival times through sociomaterial networks. *Environment and Planning B*, 35(6), 9976–1011

Schwanen, T., Dijst, M. and Dieleman, F. 2004. Policies for urban form and their impact on travel: The Netherlands Experience. *Urban Studies*, 41(3), 579–603

Schwanen, T., Ettema, D. and Timmermans, H. 2007. If you pick up the children, I'll do the groceries: Spatial differences in between-partner interaction in out-of-home household activities. *Environment and Planning A*, 39, 2754–73

Schwanen, T. and Lucas, K. 2011. Understanding auto motives, in *Auto Motives: Understanding Car Use Behaviours* edited by K. Lucas, E. Blumenberg and R. Weinberger. Bingley: Emerald, 3–39

Scottish Executive Central Research Unit 2002. Why do parents drive their children to school? Edinburgh

Scottish Household Survey 2008. Edinburgh

Sociaal Cultureel Planbureau (SCP) 1999. De stad op straat; De openbare ruimte in perspectief, SCP rapport. Rijswijk

SCP 2003. De Meerkeuzenmaatschappij; Facetten van een temporele organisatie van verplichtingen en voorzieningen, SCP, Den Haag

SCP 2004. De Veeleisende samenleving. Den Haag

SCP 2005. Anders onderweg. Den Haag

SCP 2006a. De tijd als spiegel. Hoe Nederlanders hun tijd besteden. Den Haag

SCP 2006b. Angstige Burgers; de determinanten van de gevoelens van onveiligheid onderzocht. Den Haag

SCP 2007. Nieuwe links in het gezin. Den Haag

SCP 2008. De ontwikkeling van de AWBZ-uitgaven. Den Haag

SCP 2009a. De toekomst van de mantelzorg. Den Haag

SCP 2009b. Kunnen alle kinderen meedoen? Onderzoek naar de maatschappelijke participatie van arme kinderen. Den Haag

SCP 2010. Tijd op orde? Een analyse van de tijdsorde vanuit het perspectief van de burger. Den Haag

Sennett, R. 1976. *The Fall of Public Man*. New York: Random House

Sennett, R. 1998. *The Corrosion of Character*. New York: Norton

Sennett, R. 2006. *The Culture of the New Capitalism*. New Haven: Yale University Press

Shafer, A. and Victor, D. 2000. The future mobility of the world population. *Transportation Research*, 34, 171–205

Shared Space 2006. Room for everyone: A new view on public space, edited by Herngreen, R. Groningen: Keuning Institute

Shell Deutschland 2009. Shell PKW-Szenarien bis 2030; Fakten, trends und Handlungsoptionen fur nachhaltige Auto – Mobilitat

Shell International 2008. *Shell Energy Scenarios to 2050*

Shell 2011. *Signals and Signposts: Shell Energy Scenarios to 2050*

Sheller, M. 2003. Automotive emotions: Feeling the car. *Theory, Culture and Society*, 21, 221–42

Sheller, M. 2004. Mobile publics: beyond the network perspective. *Environment and Planning D: Society and Space*, 22, 39–52

Sheller, M. and Urry, J. 2006. The new mobilities paradigm. *Environment and Planning A*, 38, 207–26

Shove, E. 2002a. Converging conventions of comfort, cleanliness and convenience. *Journal of Consumer Policy*, 26(4) 395–418

Shove, E. 2002b. Rushing around: Coordination, mobility and inequality. Draft paper for the Mobile Network meeting, October

Shove, E. and Walker, H. 2007. Caution! Transition Ahead: Politics, practice and sustainable transition management. *Environment and Planning A*, 39(4), 763–70

SIKA (Swedish Institute for Transport and Communication Analysis) 2006. RES 2005–2006 The National Travel Survey

Simmons, M. 2008. Global crude oil supply: is the oil peak near? *World Energy Magazine*

Skinner, C. 2003. Running around in circles: How parents coordinate childcare, education and work. London: Joseph Rowntree Foundation

Skinner, C. 2005. Coordination points: A hidden factor in reconciling work and family life. *Journal of Social Policy*, 34(1), 99–119

Smith, N. et al. 2006. Evidence base review on mobility: Choices and barriers for different social groups, Paper prepared for the Ministry of Transport, UK, Centre for Research in Social Policy

Snellen, D. 2001. Urban form and activity patterns: An activity-based approach to travel in a spatial context. Thesis, Technical University Eindhoven

Snellen, D., Hilbers, H. and Hendriks, A. 2005. Nieuwbouw in beweging. Een analyse van het ruimtelijk mobiliteitsbeleid van Vinex, Ruimtelijk Planbureau. Den Haag: NAi,

Social Exclusion Unit 2003. Making the connections: Transport and social exclusion, London

Solomon, J. and Titheridge, H. 2009. Setting accessibility standards for social inclusion: Some problems. UTSG, January, London

Soron, D. 2009. Driven to drive: Cars and the problem of 'Compulsory Consumption', in *Car Troubles* edited by J. Conley and A. Tiger MacLaren. Aldershot: Ashgate, 181–97

Southerton, D. et al. 2001. The social worlds of caravanning: Objects, scripts and practices, *Sociological Research Online*, 6(2)

Southerton, D. 2003. Squeezing time: Allocating practices, coordinating networks and scheduling society. *Time & Society*, 12(1), 5–25

Sperling, D. and Clausen, E. 2003. The developing world's motorization challenge. *Issues in Science and Technology*, 19(1), 52–9

Sperling, D. and Gordon, D. 2008. *Two Billion Cars: Driving towards Sustainability*. Oxford: Oxford University Press

Sperling, D. 2010. Steps into postfossil mobility: A vision and policy plan for sustainable transportation. Keynote lecture. 'Future Technologies II: Mobility', Our Common Future, Essen, 4 November

Staal, P.E. 2003. Automobilisme in Nederland: een geschiedenis van gebruik, misbruik en nut, Zutphen: Walburg Pers

Staley, S. 2006. The sprawling of America ; In defense of the dynamic city, Los Angeles, Reason Foundation

Staley, S. 2009. *Mobility First: A New Vision for Transportation in a Globally Competitive 21st Century*. Lanham, MD: Rowman & Littlefield Publishers

Steer Davies Gleave 2005. *Overcoming Car Dependence*, Transport for London

Steg, L. 2005. Car use: Lust and must. Instrumental, symbolic and affective motives for car use. *Transportation Research Part A: Policy and Practice*, 147–62

Steg, L. and Gifford, R. 2005. Sustainable transportation and quality of life. *Journal of Transport Geography*, 13, 59–69

Stern, N. 2006. *The Economics of Climate Change*. Cambridge: Cambridge University Press

Stokes, G. 2002. Reducing reliance on the car in rural areas. Paper, ETC

Stopher, P. and Fitzgerald, C. 2008. Managing congestion – are we willing to pay the price? Working Paper Institute of Transport and Logistics, University of *Sydney*

Stradling, S. 2002. Reducing car dependence, in *Integrated Futures and Transport Choices*, edited by J. Hine and J. and Preston. Aldershot: Ashgate, 100–115

Stradling, S. Noble, A. et al. 2002. Eight reasons people don't like buses. Paper, Napier University. Edinburgh

Stradling, S. 2007. Determinants of car dependence, in *Threats from Car Traffic to the Quality of Urban Life*, edited by T. Garling and L. Steg. New York: Elsevier, 187–204

Sturm, R. 2005. Child obesity – What can we learn from existing data on societal trends? *Public Health Research, Practice and Policy*, 2(2) 1–9

Stutzer, A. and Frey, B. 2004. Stress that doesn't pay: the commuting paradox. *Working Paper No 151, Institute for Empirical Research in Economics*, University of Zurich

Surface Transportation Policy Project 2003. Transportation Costs and the American Dream

Surface Transportation Policy Project 2005. Driven to Spend: Pumping Dollars out of our Households and Communities

Tacken, M. no date. Ouderen en hun mobiliteit buitenshuis, problemen en alternatieven

Taskforce Mobiliteitsmanagement 2008. Voorstel Taskforce (voorzitter: L. de Waal)

Taskforce (her) ontwikkeling bedrijventerreinen 2009. Kansen voor kwaliteit ; een ontwikkelingsstrategie voor bedrijventerreinen (voorzitter; P. Noordanus)

Taylor, J. et al. 2009. The travel choices and needs of low income households: The role of the car. National Centre of Social Research

Teisman, G. 2005. Publiek Management op de grens van chaos en orde: over leidinggeven en organiseren in complexiteit, Den Haag: Academic Service

Teisman, G., Buuren, van, A. and Gerrits, L. 2009. *Managing Complex Governance Systems: Dynamics, Self-Organization and Co-evolution in Public Investments*. New York: Routledge

Thorgersen, J. and Moller, B. 2007. Breaking car-use habits: The effectiveness of economic incentives. *Transportation*, 3, 329–45

Thomsen, T., Drewes, L. and Gudmundsson, H. 2005. *Social Perspectives on Mobility*. Aldershot: Ashgate

Thomson, L. 2009. How times have changed. Active transport literature review. Victoria Health Organisation

Thulin, E. and Vilhelmson, B. 2007. Mobiles everywhere: Youth, the mobile phone and changes in everyday practice. *Young*, 15(3), 235–53

Tillema, T. 2007. Road pricing: A transport geographical perspective. Thesis, Utrecht: Geografische Studies Utrecht

Titheridge, H. 2008. Social exclusion, accessibility and lone parents. Paper presented at the UK – Ireland Planning Research Conference, Belfast, 18-20/3

Todman, L. 2003. Physical mobility and social exclusion: Some preliminary thoughts on the explanatory power of the Spatial Mismatch Hypothesis. Paper presented at the Move Observatory Conference, Verona

Tranter, P. and Malone, K. 2003. Out of bounds: Insights from children to support a cultural shift towards sustainable and child-friendly cities. Paper for National Conference State of Australian Cities, RMIT University and UNSW Australian Defence Force Academy

Tranter, P. and Sharpe, S. 2007. Children and peak oil: An opportunity in crisis. *International Journal of Children's Rights*, 15, 181–97

TRB and Institute of Medicine 2005. Does the built environment influence PHYSICAL ACTIVITY?: Examining the evidence

TRB 2009. Driving and the built environment: The effects of compact development on motorized travel, energy use and CO_2 emissions

Tulloch, J. and Lupton, D. 2003. *Risk and Everyday Life*. New York: Sage

Tully, C. 2003. Young people living in a mobile world. Thesis on the cryptic socialization of mobility regarding the next generation. Conference, Alternative Mobility Futures, Lancaster University, UK

Universiteit van Hasselt, Instituut voor Mobiliteit 2008. Onderzoek Verplaatsingsgedrag Vlaanderen 3 (2007–2008)

Urry, J. 2000. Inhabiting the car. *Department of Sociology Discussion Papers*, Lancaster University, UK

Urry, J. 2000. *Sociology Beyond Societies*. London: Routledge

Urry, J. 2003. Social networks, travel and talk. *British Journal of Sociology*, 54(2), 155–75

Urry, J. 2004. The 'System' of Automobility. *Theory, Culture & Society*, 21(4/5), 25–40

Urry, J. 2006. Climate change, travel and complex futures. *British Journal of Sociology*, 59(2), 261–79

Urry, J. 2007. *Mobilities*, Cambridge: Polity Press

VanderBilt, T. 2008. *Traffic: Why We Drive the Way We Do*. New York: Alfred A. Knopf

Verburg, T., Dijst, M. and Schwanen, T. 2005. Leef- en mobiliteitsstijlen Stedenbaan, Paper Faculteit Geowetenschappen, Utrecht

Vereniging Nederlandse Autoleasemaatschappijen 2007. Zicht op de zakenautorijder

Vigar, G. 2001. *The Politics of Mobility*. New York: Spon Press

Vilhelmson, B. 2005. Urbanisation and everyday mobility. Long-term changes of travel in urban areas of Sweden. *Cybergeo: Revue europeenne de geographie*, no 302, 17 Fevrier

Vilhelmson, B. 2007. The use of the car: Mobility dependencies of urban everyday life, in, *Threats from Car Traffic to the Quality of Urban Life*, edited by T. Garling and L. Steg. New York: Elsevier, 145–64

Vinne, V. van der. 2010. De autoproblematiek in Nederland, Zutphen, uitgeverij Siemes

Virilio, P. 1997. *The Open Sky*. London: Verso

Vleugel, J. 2009. Elektrische autos: vooruitgang of stilstand ? *Bijdrage Colloquim Vervoersplanologische Studiedagen*, Antwerpen, 19–20 November

VNG 2009. THB-advies goed uitgangspunt voor verder overleg, VNG, Den Haag

Voss, J. 2003. Shaping socio-ecological transformation: The case for innovation governance. Paper presented at the Open Science meeting of the International Human Dimensions Programme of Global Environmental Change Research, Montreal

Voss, J. et al. 2006. Steering for sustainable development: A typology of empirical contexts and theories based on ambivalence, uncertainty and distributed power. Paper for workshop Governance for Sustainable Development, 6–7 February, Berlin

Voss, J., Bauknecht, D. and Kemp, R. 2006. *Reflexive Governance for Sustainable Development*. Cheltenham: Edward Elgar

VROM Raad 2006. Werklandschappen ; een regionale strategie voor bedrijventerreinen, advies 053, Den Haag

Wajcman, J. 2008. Life in the fast lane? Towards a sociology of technology and time. *The British Journal of Sociology*, 59(1), 59–77

Warde, A. et al. 2004. The changing organization of everyday life in UK: evidence from time use surveys 1975–2000. Paper presented at the Time Use Seminar, CASS Business School, London

Watson, M. and Shove, E. 2005. Doing it yourself? Products, competence and meaning in the practices of DIY. Paper, European Sociological Association, Torun

Watson, M. and Shove, E. 2006. Materialising consumption: Products, projects and the dynamics of practice, Paper in Cultures of Consumption Research Programme, Birkbeck College, London

Webber, M. 1992. *The Joys of Automobility*. Reprint 110, Transportation Center, University of California, Berkeley

Wee, van, B. 2012. *Transport and Ethics: Ethics and The Evaluation of Transport Policies and Projects*. Cheltenham: Edward Elgar Publishing

Wee, van, B. and Chorus, C. 2009. Accessibility and ICT: A review of literature, a conceptual model and a research agenda. Paper presented at the BIVEC-GIBET research day, 27 May, Brussels

Weider, M. 2004. China – Automobilmarkt der Zukunft ? Berlin, WZB discussion paper

Weiss, M. and Heywood, J., et al. 2000. On the road in 2020: A life-cycle analysis of new automobile technologies. Energy Laboratory Report, Cambridge: MIT.

Wells, P. 2006. Alternative business models for a sustainable automotive industry. Proceedings, Changes to Sustainable Consumption Conference, 20–21 April, Copenhagen

Weston, L. 2005. What helps and hinders the independent travel of non-driving teens? Thesis, University of Austin, Texas

Weston, L. and Handy, S. 2005. Driving by choice or necessity? Research Report SWUTC/05/167522-1, University of Texas, Austin

Whitmarsh, L. and Kohler, J. 2010. Climate change and cars in the EU: The roles of auto firms, consumers and policy in responding to global environmental change. *Cambridge Journal of Regions, Economy and Society*, 427–41

Williams, F. et al. 2001. Consumption, exclusion and emotion: The social geographies of shopping. *Social & Cultural Geography*, 2(2), 203–20

Williams, J. and Green, G. 2001. Literature review of public space and local environments, Oxford Centre for Sustainable Development

Woodcock, J. and Aldred, R. 2008. Cars, corporations and commodities: Consequences for the social determinants of health. *Emerging Themes in Epidemiology* 5: 4

World Business Council for Sustainable Development 2004. Mobility 2030: Meeting the challenges to sustainability

Wright, B. 2008. No way to go: A review of the literature on transportation barriers in health care. *World Transport Policy & Practice*, 14(3), 7–23

Zandvliet, R. 2005. In perpetual motion. Time–space variations in the characteristics of visitor populations and the performance of places. Thesis, University of Utrecht

Zandvliet, R., Dijst, M. and Bertolini, L. 2002. Diurnal Variations in visitor populations from a transportation perspective. Paper for 82nd meeting of TRB

Zhang, M. 2006. Travel choice with no alternative: Can land use reduce automobile dependence? *Journal of Planning Education and Research*, 25, 311–26

Zhu, Y. 2005. Energy and motorization: Scenarios for China's energy, 2005–2020, WZB Discussion Paper, Berlin

Zinn, J. 2006. 'Biographical Certainty'. Reflexive modernity, SCARR, Working Paper 2006/ 15

Zumkeller, D., Chlond, B. and Ottmann, P. 2005. Car dependency on household and personal level: Transitions of car ownership and future development of motorization in Germany, based on the German Mobility Panel (MOP), Karlsruhe

Zumkeller, D. 2003. Fordert Telekommunikation den Bedeutungslust der Nahe?; ein Zukunftsbild unserer Mobilitat, Vortag Tagung City.net – Stadte im Zeitalter der Telekommunikation, 19 Juni, Weimar

Zwart, W van der. 2007. De wereld als spiegel: Identiteitsvorming in een veranderende samenleving, Radboud Universiteit Nijmegen

Zwerts, E and Werts, G 2006. Children's travel behaviour: A world of difference, Hasselt University. Paper for TRB

Index